THE
WONDERFUL
MR WILLUGHBY

THE
WONDERFUL
MR WILLUGHBY

The First True Ornithologist

TIM BIRKHEAD

BLOOMSBURY PUBLISHING
LONDON • OXFORD • NEW YORK • NEW DELHI • SYDNEY

BLOOMSBURY PUBLISHING
Bloomsbury Publishing Plc
50 Bedford Square, London, WC1B 3DP, UK

BLOOMSBURY, BLOOMSBURY PUBLISHING and the Diana logo are trademarks of
Bloomsbury Publishing Plc

First published in Great Britain 2018

Copyright © Tim Birkhead, 2018
Maps by Martin Lubikowski
Family tree and diagrams by Phillip Beresford

Tim Birkhead has asserted his right under the Copyright, Designs and Patents Act, 1988,
to be identified as Author of this work

For legal purposes the Acknowledgements on p. 273
constitute an extension of this copyright page

A catalogue record for this book is available from the British Library

Library of Congress Cataloguing-in-Publication data has been applied for

ISBN: HB: 978-1-4088-7848-4; EPUB: 978-1-4088-7850-7

2 4 6 8 10 9 7 5 3 1

Typeset by Newgen KnowledgeWorks Pvt. Ltd., Chennai, India
Printed and bound in Great Britain by CPI Group (UK) Ltd, Croydon CR0 4YY

To find out more about our authors and books visit www.bloomsbury.com
and sign up for our newsletters.

Contents

Preface

He digested the whole history of nature with that spirit and judgement
that it always appeared new; with that care and diligence that he was always
constant to himself; and with that integrity that he has ever been esteemed
a faithfull interpreter of nature
                    Memorial to Francis Willughby in Middleton Church

Until very recently Francis Willughby was a little-known seventeenth-century natural historian overshadowed by his brilliant
Cambridge tutor, friend and collaborator, John Ray. Having barely
begun his scientific career, Willughby died in 1672 at the age of
just thirty-six. Ray, through his own much longer lifetime, brought
Willughby's 'works' to fruition, and made sure they were published. Paradoxically, this selfless act resulted in Ray unwittingly
monopolising the accolades – not without some justification – for
what was very much a joint effort. The widespread perception
of Ray's intellectual superiority over Willughby is largely a consequence of an adulatory biography of Ray dating from 1942, by
Canon Charles Raven, whose motivation, in part at least, was to
correct an earlier perception that Willughby was the true genius of
the two. However, when Raven's book was 'in press' he learned that
there existed an extensive Willughby family archive. Realising what
he might have missed, he was able to do little more than add a note
to the proofs saying: 'Some day a book on him & his collection
must certainly be written.' The Middleton Collection as it is now

known, lodged in the Nottingham University Library, provided much of the material for the present book, and now allows us to paint a far more complete portrait of Willughby's short life. Some sense of the collection's richness can be gleaned from the fact that after being invited to view Willughby's bird paintings and pressed plants in April 1942, after his book was published, Raven wrote to thank Lord Middleton and to tell him that it was one of the most thrilling experiences of his working life.

My own involvement with Francis Willughby was the result of assuming Canon Raven's persuasively detailed case to be correct and that Ray was the intellectual *tour de force* behind his and Willughby's published works. I had gone to the Willoughby* family home to photograph their portrait of Francis for a book I was writing, whose title, *The Wisdom of Birds*, and content, celebrated the contribution that John Ray had made to ornithology, in part through his book *The Wisdom of God* (1691). After I had photographed the portrait, the elderly Lord Middleton (Francis Willughby's descendant) and his wife invited me for lunch. During our conversation I commented on Ray's brilliance, partly because it was true, but also because I knew so little about Francis Willughby. Lady Middleton's reaction was a rebuke of such ferocity that it took my breath away. Her husband, embarrassed I suspect, said nothing. Lady Middleton made it very clear to me that the genius was Willughby rather than Ray, who, she said, was a mere servant. What I did not realise at the time, as will become apparent, was that by attributing the success of their joint ventures to Ray I had reopened a deep wound inflicted some three and a half centuries earlier. Hastily, I moved the conversation on, but as I drove home later that day, I reflected on Lady Middleton's reproach and in a flash it came to me that Francis Willughby was in need of some attention.

Soon after starting my investigations it became clear why Willughby has been so neglected by those interested in the history

---

* Francis Willughby spelt his name as here, but previous and subsequent generations of his family spelt it 'Willoughby' (Cram et al. 2003:1).

of science: most of his original notes and letters have been lost. This loss is a consequence of Willughby dying before his time and before his talents had been fully recognised; but also – paradoxically – it is due to the efforts of his friends and family, when they eventually realised his significance, as they exchanged vital documents in an effort to preserve Willughby's memory. In contrast, John Ray lived long enough to become famous during his own lifetime, while his colleagues took care to preserve his correspondence and notes after his death. Ray was a prolific, diligent letter-writer and note-keeper, and by publishing much of his own work he guaranteed his efforts a permanence that Willughby's lacked. On top of this, the lives of the two men were intimately intertwined: they studied together, travelled together, discussed ideas together, and lived and worked under the same roof for years. This means that while the writings of John Ray provide a wealth of material about Willughby, we have little choice but to view the latter largely through Ray's eyes. Given the priority that the two men gave to objectivity in the way they conducted their science, it is unfortunate that Ray's early portrait of Willughby is not an entirely neutral one.

After my visit to the Middleton home, I sought and obtained funding from the Leverhulme Trust for what they called an International Network Grant: a scheme to bring together academics from different countries to focus on a single project. The Willughby Network comprised fifteen members with expertise spanning a wide range of topics in the history of science and linguistics, with myself as the sole scientist. A major source of information for us all has been the hoard of Willughby material held by the family, which Raven was unable to use in his book. Those papers, together with other remarkable discoveries, allow us to rebuild Francis Willughby's reputation.

This is a biography of science recounted through the activities of one man and his like-minded colleagues, at the dawn of the scientific revolution. Francis Willughby's short life spanned one of the most remarkable periods of history – an age when it became clear that science or the 'new philosophy', as he and his colleagues called it, had the potential to explain the wonders of the natural world.

It has been a challenge for me to imagine a world on the brink of scientific discovery: I have had to keep checking myself from criticising Francis for knowing so little. Things that seem so blindingly obvious today were far from obvious then, and it is surprisingly difficult to throw off three and a half centuries of natural history knowledge and imagine oneself so ignorant and uncertain. One way of doing this is to draw comparisons between what we know today and what was known in the past, but herein lies the potential trap of what historians of science call 'Whiggishness', that is, interpreting past events through a modern lens. I have tried to avoid this, but as a scientist I have felt it essential to place Willughby's observations in a modern context so as to better evaluate them. I have also tried to relive some aspects of Willughby's life – including the excitement and sheer novelty of discovering the external features and internal anatomy of birds in his quest for essential identification marks. Francis's brief journey was full of excitement and, among other things, the realisation that discovery is a way of knowing. It is a pleasure I have shared, both through him, but also through my own ornithological research. There's nothing like the quest for knowledge for getting you out of bed in the morning.

Francis Willughby has been referred to as a 'virtuoso', a term used to distinguish him, and people like him, from others in the seventeenth century who were 'academics', employed at the universities of Oxford, Cambridge and elsewhere. The term 'virtuoso' reflects Willughby's status as an amateur whose social position as a member of the landed gentry freed him from the necessity of earning a living at a university. Of course, 'virtuoso' also implies unique expertise and an interest in many things: in short, a polymath.

Today's scientists, by contrast, are almost all specialists, even though they are expected to be both experts and generalists. Knowledge is now so extensive and growing at such a rate that few academics have the intellectual appetite, ability or opportunity to be polymaths. The surge in knowledge following the scientific revolution in the seventeenth century was so great that by the mid-1800s it was deemed impossible for anyone to know everything, although it was still possible to be an expert on a wide range of subjects. Francis Willughby didn't know everything of course, but he was considered to be more knowledgeable about natural history than anyone else in the mid-1600s. By a fortunate set of circumstances he found himself at the right place at the right time surrounded by the right group of people engaged in an entirely novel way of looking at nature. Willughby was there at what I consider to be the inception of modern science, and became one of its most interesting and impressive proponents.

He didn't act alone. In the 1600s for the first time there was a community – a small one – of would-be scientists and natural philosophers in Britain and on the continent who could exchange ideas, challenge each other, and use their combined knowledge to generate new ideas and better understanding. This tiny community eventually helped to form the Royal Society, whose motto, *Nullius in verba* or 'Take nobody's word for it', captured the spirit of the endeavour, enabling the members to cast aside the mantle of Ancient Greek knowledge. The fledgling Royal Society, of which Francis was an 'original member', provided the beginnings of a scientific infrastructure that would foster science in future generations. The Royal Society provided a meeting-place for those with similar interests; it offered demonstrations of scientific or natural history phenomena, and, perhaps most important of all, it provided a place where the new philosophers could publish – and hence share – their findings with a wider audience. It still does. Francis Willughby's innovative work and industry inspired three major natural history volumes. One on birds: *Francisci Willughbei: Ornithologiae Libri Tres* (1676), written in Latin, whose English translation, *The Ornithology of Francis Willughby* (1678), I shall simply refer to as the *Ornithology*; one on fishes, *Historia Piscium* (1686), and one on insects, *Historia Insectorum* (1710), both written in Latin, but which for simplicity I shall occasionally refer to as the *History of Fishes* and the *History of Insects*, respectively. These formed the foundation of a new type of natural history, with the study of birds being the most significant.

Given Francis's varied interests and achievements, why have I focused on his success as an ornithologist? There are three reasons. First, birds were the initial focus of his efforts; second, birds set a model for how his studies of other animals, such as fish and insects, would subsequently be organised; and third, of his works – all published after his death – that on birds, while not well known, is certainly the best known, and by far the most interesting. This is the story of the man who began the scientific study of birds.

I

Bitten by the Snake of Learning

The sky is a clear speedwell blue. It is hot, almost 35°C, but there's a welcome westerly breeze. Earlier in the week Francis Willughby and his servant, whose name we don't know, rode through the foothills of the Pyrenees at Banyuls-sur-Mer on the Mediterranean coast of France and into Spain. They are now heading southwards on mule-back towards Valencia and to avoid the heat and bandits they travel mostly in the late afternoon and at night. Willughby – twenty-eight years old – is eagerly anticipating the forbidden territory that is seventeenth-century Spain. The country is not recommended. Yet this is precisely why Willughby is here. He's testing new terrain and himself in the hope of new discoveries.

For the past eighteen months he and three friends – who have either returned home or remained behind in France – have enjoyed and endured a grand tour of Europe, travelling variously by horse, barge and cart through the Low Countries, central Europe, Italy and France. It has been an extraordinary expedition whose aim – amply fulfilled – was the acquisition of new knowledge in natural history, language, industry and indeed anything else that took their fancy. These were men on a mission. The 1650s and 1660s marked the beginning of the scientific revolution and in their quest for information Willughby and his friends had travelled hard, with little time for relaxation. They visited other scientists, savants and philosophers to scrutinise their cabinets of curiosities, purchase specimens and illustrations, dissect birds and fish, and most significantly, to take notes – lots of notes – recording everything they saw. It was a journey that changed them, and it would change ornithology as well.

The road to Valencia, little more than a track, is dry and dusty, the air redolent with the scent of grass scorched by a relentless sun. On either side, the hills are covered by a sea of yellow star-thistle and other unfriendly vegetation. Blue-winged grasshoppers explode from under the mules' hooves, and the hot air rattles with cicadas. Looking up from under the brim of his hat, Francis is surprised by a sky full of birds. Reining in his mule, he stops and, shielding his eyes with his hand, watches in awe as hundreds of birds trace swirls of interlocking spirals as the sunlight creates golden windows in their dark-edged wings. Obviously raptors, but what are they: puttocks? gleads? more-buzzards? honey-buzzards? Because the birds are so far away, it is hard to tell. Francis watches as they gain height, and then, almost as though at a signal, the tiny silhouettes cease circling and on stiff wings they glide swiftly away one after the other to the south.

They were honey-buzzards. Pouring out from their central European breeding grounds and heading towards the Strait of Gibraltar where they would cross the Mediterranean and continue onwards to their winter quarters in central Africa. But Francis did not and could not have known what they were nor where they were headed.

He knew the species, however, for it was he who, a few years previously, had examined a honey-buzzard in the hand and recognised that it differed from the common buzzard. And it was Willughby who made the first accurate description of this new species.

Little could Francis Willughby have imagined that far in the future, his brief lifetime of ornithological research – encapsulated in a ground-breaking encyclopedia, known now as the *Ornithology* – would be rewarded with his name being tied to this particular bird.

As a boy, Francis probably had little interest in birds or any other aspect of natural history, except perhaps as objects to hunt. Later, at university, however, a remarkable set of circumstances was to awaken a passion that resulted in him becoming the most accomplished naturalist of his day.

The family's ancestry can be traced back to the early thirteenth century when Ralphe Bugge, a wool merchant, purchased land near the tiny village of Willoughby in the Nottinghamshire Wolds. Adopting the name, the family became 'de Willoughby' and eventually simply Willoughby, or in Francis's case, the distinctive 'Willughby'. Through careful marriages and financial management the family accrued land and wealth such that by the late 1500s they had estates in over twenty English counties: they were landed gentry.

Francis was born 'on Sunday about six of the clocke in the morning being the two and twentieth of November 1635 at the family seat of Middleton Hall, Warwickshire'.[1] He had two older sisters, Lettice, born 17 March 1627, and Katherine, who was born 4 November 1630. Francis was a family name, providing plenty of opportunity for confusion among those later researchers concerned with Willoughby history. Our Francis's father is referred to as Sir Francis, although there were other Francis Willoughbys, both earlier and later, with that same title. Middleton Hall – first mentioned in the Doomsday Book in 1086 – had been in the family for several generations, but Francis's parents were the first to make it their home. The principal seat of the Willoughby family was near Nottingham, at Wollaton Hall, a monumental property constructed at great expense during the 1580s, by our Francis's great-grandfather, another Sir Francis Willoughby, known as 'Francis the Builder'.

In the late 1500s the income of the Willoughby family came mainly from coal, iron ore and the blue textile dye, woad. Wollaton Hall – said to be the 'architectural sensation of its age'[2] – was an extravagance too far that, combined with Francis the Builder's lavish lifestyle and some failed speculations, plunged the family into debt, until they were forced to live in what was euphemistically referred to as 'financial embarrassment'.[3] Despite his evident ability to erect an enormous architectural artefact, Francis the Builder was unable to produce a male heir. Of a total of seven children, from two marriages, his only son died as an infant. Sir Francis therefore arranged for his daughter Bridget to marry a distant cousin, Sir

Percival Willoughby, thereby safeguarding what little was left of the family's fortune.

It was Sir Percival's son, Sir Francis (our Francis's father), together with his wife Cassandra Ridgeway, whose careful management of the estates got the Willoughby family back onto a firm financial footing. Betrothed in 1610, Sir Francis and his young wife – she was just fifteen – took up residence in the relatively modest Middleton Hall, north Warwickshire, in 1615. Middleton was a gift from Sir Percival, and, fortuitously, it allowed the couple to maintain a low profile during the political troubles. After Bridget's husband, Sir Percival, died in 1643, she joined her son at Middleton.[4]

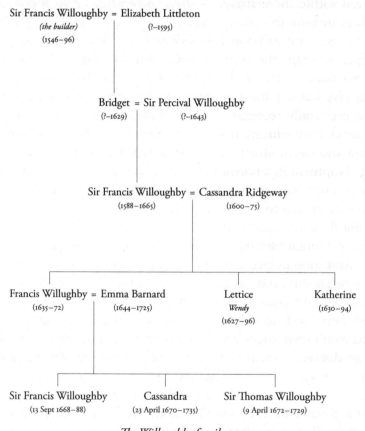

The Willoughby family tree.

So it was that our Francis emerged in the elegant half-timbered Hall at Middleton. Precious little is known of his childhood, other than that his formal education began at the free school at nearby Sutton Coldfield, where he was taught by William Hill, previously a Fellow at Merton College, Oxford, and well known for his abilities in Greek, Latin and physics. Francis and William must have started at Sutton Coldfield School at about the same time, when Francis was five or six years old, for after the death of his wife in 1641 (when Francis was six), Hill moved to London to relaunch his career as a doctor of medicine. His second marriage, to the daughter of a physician, was marred by the scandal associated with the birth of a son just seven months later, whom Hill claimed was conceived within the marriage[5] – which is not impossible since some babies are born this prematurely.

It is frustrating that we know so little of how or in what way Hill helped to shape the young Francis. The fact that in her account of her father's life, our Francis's daughter Cassandra commented that Hill was the 'most famous schoolmaster of his time' suggests that the family recognised the importance of Hill's influence on Francis's early education – presumably in his proficiency in both Latin and Greek. The value that Francis's parents placed on learning is captured in a portrait of his mother where she appears holding an open book in an attitude identical to one in which her son was later portrayed.[6]

For the landed gentry education was de rigueur, at least for their sons. Although they had no need of a profession as such, a university education helped to ensure that young men acquired 'acceptable behaviour' and became 'rationall and graceful speakers'.[7] In addition, and crucially, familiarity with the law meant that they were better able to defend themselves against legal claims associated with inheritances. Such claims were common at the time and, as we shall see, were to dominate the last few years of Francis's life.

⊦⋎⊦

Throughout Francis's schooldays the country was engulfed in the English Civil War. The Willoughbys were Royalist sympathisers, and somehow managed to stay out of trouble – and indeed, they were one of the few Royalist families to survive unscathed. Nonetheless, it must have been an extraordinarily tense time as the Parliamentarians ruthlessly plundered Royalist estates. The once-unified gentry community was divided by loyalties and disloyalties, trust and distrust. Information circulated via pamphlets, but as is so common under such circumstances, no one knew who, or what, to believe. In addition to the pitched battles, killing was random and rife in the countryside, and disease widespread. The death toll was extraordinary. In England around 85,000 people were killed, and a further 100,000 died of disease, out of a population of just five million.

Some three years after Charles I was beheaded, and with England under Parliamentary rule, the sixteen-year-old Francis entered Trinity College, Cambridge, in September 1652. Why the family decided to send him there rather than to Magdalen College, Oxford, where his father had studied, is not clear. It could have been that by then Trinity was considered the better institution. It may also have been because of the family's political sympathies. The fact that Trinity retained a renowned Royalist tutor – James Duport – may also have made that college especially attractive. Whatever the reason, Trinity could hardly have been a better choice.

Admission to Cambridge in the seventeenth century was organised hierarchically – based on wealth – under a number of different headings. At the lowest level were the sons of poor parents, known as sub-sizars, who paid no fees but earned their keep either by assisting in the college kitchen or by undertaking menial tasks for wealthier students, such as cleaning or serving food. Above them were sizars, less poor students who received financial assistance either in the form of reduced fees or the cost of their lodgings. The next level above, pensioners, paid for their own tuition and keep (referred to as 'commons'). They were a notch below the Fellow-commoners, who in turn were placed below the sons of noblemen. All of these were students inasmuch as they each had a tutor, who in turn was either one of Trinity's sixty Fellows or, very occasionally, the Master himself. Fellows were those who had been students, won a scholarship in either their second or third year of study, completing their Batchelor of Arts (BA) in the fourth and final year of their course and then were selected after being examined by the Master and senior fellows. The eight most senior Fellows were simply those who had been around the longest, and it was they who effectively ran the college. Most Fellows expected eventually to become clergymen, leaving Trinity and permitting those below them to move up the social ladder.[8]

Trinity's hierarchy was reinforced, and is apparent in numerous ways, including the so-called buttery books, in which each student's weekly food bill was logged. Financially secure if not obviously rich, Francis entered Trinity as a Fellow-commoner. This meant that he, or rather his father, paid double tuition fees, permitting Francis certain privileges, including dining with the Fellows – hence Fellow-commoner. During one of his first weeks at Trinity, Francis and the other eleven Fellow-commoners each spent just over seven shillings on meals, whereas the other students spent about one-third of this, reflecting the difference in the quality and quantity of food they consumed.

Most of those entering Trinity as either Fellow-commoners or noblemen were not interested in academic success or in acquiring a degree, since their family's wealth precluded the need for a profession. In contrast, less privileged students often worked hard to succeed and secure a profession. Isaac Barrow, for example, just a few years older than Francis, who became his friend – and was later famous for his development of calculus – started as a pensioner at Trinity. John Ray, the son of a blacksmith, who later became Francis's tutor, mentor and lifelong friend, was admitted to Trinity in 1644 as a sizar. In 1661, a decade after Francis Willughby, Isaac Newton – arguably the greatest physical scientist ever – started as a sub-sizar at Trinity.[9]

By the time Francis entered Trinity as an undergraduate, progression through the college hierarchy had been disrupted by the ejection of several Royalist Fellows in 1645 as a result of the political unrest. To maintain the statutory sixty Fellows, the college had been forced to import Fellows from other colleges. Another consequence of this disruption was that Trinity was much more relaxed about how long Fellows could remain at the university before entering holy orders, for example, and moving on.

In Willughby's day Trinity College was – and still is – one of the foremost, largest and wealthiest of the Cambridge colleges. Founded in 1546 as one of the last acts of Henry VIII, and funded by monies resulting from his dissolution of the monasteries, Trinity's magnificent buildings were constructed in the sixteenth and seventeenth centuries. The college as we see it today, with its elegant towers, chimneys and cloistered walkways, looks, with few exceptions, much as Francis Willughby experienced it.

On his arrival from Middleton, Willughby entered the college from Trinity Street, through the Great Gate, emerging into the Great Court, a huge quadrangle – breathtaking in its expansiveness – bordered by honey-coloured buildings, at the centre of which spouts the ornate fountain from which the college drew its water. Through an archway on the western side of the court, entered via a smaller quadrangle, was Nevile's Court overlooking the river Cam along which ponies pulled barges laden with goods. Thomas Nevile had been Trinity's Master in the 1590s and was responsible for much of the college's beautiful design.

Retracing his steps, Willughby passed the Master's House – at this time John Arrowsmith – towards the medieval clock tower. Adjacent to that tower was the chapel where Willughby among other students would assemble at dawn each day for prayers, and beyond which lay the Fellows' bowling green.

Since at least medieval times education and religion had been inextricably intertwined for the simple reason that 'superior erudition was the church's vital weapon'.[10] Scholars were revered and valued by both royal rulers and the Church for a variety of reasons. Henry VIII used university-trained men to justify both his divorce from Catherine of Aragon and his breach with the Catholic Church. It is hardly surprising then that in the sixteenth and early seventeenth centuries the main role of England's two universities, Oxford and

Trinity College Cambridge in the 1600s.

Cambridge, was to defend and reinforce Anglican values through the training of intellectually astute clergymen-cum-politicians.

In addition to divinity, however, a university education was essential for those anticipating a career in either medicine or the law. Starting in Elizabethan times, as the strategic value of educated men became increasingly apparent, the number of students attending university increased and continued to do so until the start of the English Civil War. During the sixteenth and seventeenth centuries, as now, universities trained undergraduates to think, to argue and to make a case – to become philosophers. And it did so through an arts-orientated curriculum comprising Greek, Latin, history, drama and poetry, but also logic, rhetoric, moral and natural philosophy.

Undergraduate teaching consisted of three broad areas of study, corresponding to three possible professions – divinity, medicine and law – that students might aspire to. First were studies of the sacred texts of Christian religion, essential for defending and upholding the Anglican Church. Second were the writings of classical antiquity – considered then to provide the foundation of all human knowledge. For the study of medicine one needed to understand the writings of ancient Greek and Arabic scholars and, by reading the works of Aristotle, Discorides, Pliny the Elder and others, students were exposed to astronomy, geometry and natural history. The writings of the ancients also provided moral guidance and aesthetic inspiration. Finally, the study of the historical accumulation of legal rights was considered essential for those concerned with safeguarding their inheritance or entering the legal profession.[11]

As was typical of his social class, Francis Willughby appears not to have anticipated entering any of these professions. However, unusually for someone of his position, he made full use of his status as a Fellow-commoner to develop and foster his academic abilities.

Even before the English Civil War the Puritans had made clear their dislike of scholarship for scholarship's sake. By the interregnum – 1649 to 1660 – their views had become more extreme and they aimed to transform the universities of Oxford and Cambridge into instruments of godly learning and teaching – to the despair of many academics, but especially those with Royalist sympathies. Fearing the demise of educational standards, one was prompted to comment that 'The garland has been torn from the Head of learning and placed on the dull brows of Disloyal Ignorance.'[12] The Puritans' narrow focus on saving souls through their preaching ministry drove academics away from divinity. Walter Charleton – a physician who later made a small contribution to the study of birds – captured the sentiment in 1657:

> Our late Warrs and Schisms, having almost wholly discouraged men from the study of Theologie; and brought the Civil Law into contempt: The major part of young Schollers in our Universities addict themselves to Physick [medicine]; and how much that conducteth to real and solid knowledge, and what singular advantages it hath above other studies, in making men true Philosophers; I need not intimate to you, who have so long tasted on that benefit.[13]

The Puritans' active encouragement of alternative Christian denominations and the resulting breakdown of the Church of England provided even less incentive for anyone to train for the ministry. With no need to produce Anglican clergy, the universities found themselves with a crisis of purpose. The situation was especially difficult for the older, established and more religious academics, but for those just entering university like Francis, this crisis became an opportunity since its main effect was to shift the focus away from divinity towards law and in particular towards medicine.

When Willughby entered Cambridge in 1652, education was changing and there was an accelerating interest in what was known as 'the new philosophy'. The old philosophy was based largely on the unquestioning assumption that God dictated man's place in nature, and on the authoritative knowledge of Ancient Greek authors such as Aristotle. Paradoxically, increasingly detailed study of the ancients' armchair writings merely identified the shortcomings and limitations of their approach rather than reinforcing its utility. The new philosophy sprouted from the seeds of doubt sown by the critical study of the old philosophy.

Described as the 'reasoned knowledge of nature', the new philosophy, or new science as I shall call it, had its beginnings in the thinking and writings of Francis Bacon. Staggeringly precocious, Bacon had entered Trinity College, Cambridge, in 1573 at the tender age of twelve. As he matured he favoured Renaissance ideas over what he considered the stultifying and unprogressive views of Aristotle and other ancients, and strove to create a new philosophy of science based on hypotheses and experiments. As he said, his approach to science would 'spark a light in nature that would eventually disclose and bring into sight all that is most hidden and secret in the universe'.[14] Bacon's ultimate aim was to create new, objective knowledge that would serve mankind. Christopher Columbus's discovery of North America in 1492 inspired Bacon because more than anything else it demonstrated unequivocally that there was new knowledge to be had. Columbus had not only found an unknown continent, but also – eventually – shed light on a veritable treasure trove of unknown animals and plants.

Until the early seventeenth century it was assumed that Aristotle and other Greek authors had documented almost everything there was to be known. Medieval scholars laboured to interpret and comment on Aristotle's extensive works and their studies served to reinforce, but certainly not challenge, its authority, nor was there any incentive to discover anything new. Scholars studied Aristotle in a quest to recover lost knowledge, and studied natural philosophy as a way of interpreting him. The existence of an unbreakable chain of authority linking religion with Aristotle is made clear by those who dared to submit that he might sometimes be wrong. The mathematician Jean Taisnier suggested as much in the 1500s, and was challenged by a representative of the Pope to prove his case.[15]

The Church vigorously resisted the idea that there might be anything new to discover. No one was expected to challenge the authority of Aristotle, religion, or even what they could see with their own eyes. For Francis Bacon the discovery of America made it obvious that there was new knowledge to be had. It was also the stimulus to seek new knowledge, strangely perhaps, for Bacon made no discoveries himself other than a method for making them.

The new philosophy was essentially the scientific revolution, whose impact was first felt in the cosmological changes implicit in Copernicus's *Revolution of the Heavenly Spheres* of 1543. This eventually grew into the cosmology that Isaac Newton's mathematical physics triumphantly endorsed in his *Principia* of 1687.[16] Pioneered by Bacon, Galileo and René Descartes, the new science was based on objective knowledge derived from personal observation and experimentation. At the same time, there was a growing awareness of the value of mathematics for understanding natural phenomena, particularly the movements of celestial bodies – exemplified by Isaac Newton's work, but also by some of his predecessors at Trinity College.

Civil turmoil created academic turmoil, with the decline in divinity eliciting an increase in the popularity of the new science. This was an exciting time to enter university. Willughby was doubly fortunate in being surrounded by a group of similarly inspired men, and by being supervised and mentored by James Duport, the most committed and successful of Cambridge tutors. Duport, whose father had been Regius Professor of Greek at Trinity, made numerous translations, including some of Aristotle's works, but considered himself mainly a poet. He remained a prolific versifier throughout his life, but the well-being and instruction of undergraduates was his true vocation.[17]

Being a college tutor brought in additional income and each academic year Duport took on around a dozen tutees, many more than his colleagues. Financial reward was secondary to the personal recompense Duport derived from educating his charges. He was popular, tutoring the sons of both Royalists and Parliamentarians without favour. However, Royalist parents probably chose Trinity because of Duport and the way he cared for his students. Meeting daily, he familiarised each new undergraduate with their course of study, directed their reading, and taught them the basics of the different disciplines. Then as now, such close supervision provided not only an extremely effective form of education but an excellent opportunity for tutors and tutees to get to know each other; friendships that in later life served to further careers.[18]

The ivory towers of academia were not immune from the troubles that terrorised much of the country during the English Civil War and, as a staunch Royalist and Anglican, Duport had been lucky to survive the Puritans' purges. In 1644 and 1645 no fewer than forty-nine of his sixty Trinity colleagues had been identified as 'malignants' (Royalist) and sacked. Duport seems to have escaped because of his 'inoffensive and amiable disposition' together with his 'talents and scholarship'.[19] Although the Puritans deprived him of his archdeaconry at Stowe in Lincolnshire in 1643, Duport was allowed to remain at Trinity and continue teaching. Following another purge in 1650, however, he was forced to resign his Greek professorship for refusing to subscribe to the 'Engagement for maintaining the Government without King or House of Peers'. So highly did Trinity value him, they made Duport – despite some 'statuable irregularity' – a Senior Fellow and allowed him to retain his residence at Trinity.

It was in this capacity that Francis Willughby came under Duport's charge.[20]

Duport's selfless commitment to his tutees included giving them a set of instructions he had written. Duport's rules provide us today with a wonderful feel both for his teaching philosophy and what he considered important for his students' education. As one commentator said, those rules suggest that Duport had 'more than a superficial understanding of the natural slothfulness and waywardness of youth enjoying a first taste of freedom away from the parental home'. For example, he urged his students to regularly attend St Mary's Church and take notes of the sermon; to come to chapel 'not drooping [sic] in (after the uncouth & ungodly manner of some)', and to avoid sleeping during prayers and sermons 'for that is the sleep of death'. He also advised them to avoid football as a source of recreation, 'it being ... a rude, boisterous exercise being fitter for Clownes than for Schollers'.[21]

More broadly, Duport advised his tutees to 'Write frequently to your Parents & Friends, to ye former especially if you know they desire you & expect it'; to 'Walk often in the fields, and to walke alone; for that put good thoughts into you, & make you retire into your self and commune with your own heart'; and perhaps most poignantly of all, to 'Think every day to be your last, and spend it accordingly.'

You cannot help but like the man.

Duport's rules of conduct were not solely for his students while at university; they were a guide to life itself and especially appropriate for those who, like Francis Willughby, continued to study long after leaving Cambridge. Judging from the account that Willughby's daughter left of her father's subsequent daily routine, it is clear that he assiduously followed his tutor's advice in keeping set hours for study.[22]

Conversely, Duport's firm advice about avoiding cards or dice may have ignited in Willughby a subliminal rebellious streak. The university's statutes were explicit: 'If anyone is detected as having even once taken part in a game of chance, let him be expelled from the College for ever.' Within a year or two of leaving university Francis had developed – as we will see later – a fascination for games of chance, albeit as a researcher rather than a gamester.

Among Duport's various suggestions for *how* to study, perhaps the most important was the keeping of a commonplace book. This was a way of collecting and organising knowledge; a sophisticated, personal scrapbook of information that was neither diary nor journal. A commonplace book served as a highly organised aide-memoire in which information was placed under sets of headings identified by the owner depending on his course of study.

Remarkably, Willughby's commonplace book has survived and provides a revealing window onto his Cambridge education. He had apparently obtained this substantial leather-bound book second-hand, for some of the handwriting indicates that it started life as a legal commonplace book. As was typical of commonplace books, Willughby introduced numerous headings relating to his reading under which he then added quotations, notes and references to his sources.

Despite their essential role in helping scholars to organise their knowledge, a major limitation of the commonplace book was, as anyone who has kept a notebook knows, its inflexibility. As pages on particular topics fill up, new pages, not necessarily continuous, need to be started, creating a fragmented account. To deal with this, Willughby later produced an index for himself on a separate sheet of paper.

Another improvement, suggested by the ardently Royalist scholar Thomas Harrison in the 1640s, was to keep notes on separate slips of paper (anticipating the twentieth-century index-card system) that could be filed together. Despite being hailed by some, including the polymath Samuel Hartlib, as a major advance in the organisation of information, Harrison's method was one that few appeared to follow up, almost certainly because the slips of paper lacked the permanence and convenience of a bound notebook.[23]

Seeing Willughby's commonplace book in the Middleton Collection at the University of Nottingham for the first time, I was surprised by its sheer size. With around 300 A4-sized pages and weighing 1.5 kilograms, this was a book for someone seriously interested in taking notes. The brown leather cover possesses a beautiful 350-year-old patina, while inside, Francis's distinctive handwriting – in dark and sometimes pale brown ink distilled from oak galls – spiders its way across the pages in hurried enthusiasm. Just as I do when testing a ballpoint pen for my own notebooks, Willughby had tested his quill inside the back cover, creating a pattern of random scribbles resembling those on the eggs of certain birds.

Much of what Willughby has written looks utterly illegible, and is not helped, in my case, by being mostly in Latin. My colleague Richard Serjeantson – a Trinity Fellow himself – is proficient in both Latin and the deciphering of illegible hands, and has been able to translate much of Francis's scrawl. As he so aptly said, the handwriting 'consistently gives the impression of someone who does not wish you to suppose that he had labored over his penmanship'.[24] In some ways this is not surprising: this was Willughby's notebook, for his use alone, and so as long as he could read it, that was sufficient.

Not only is the handwriting difficult to interpret, so too is the content, for the information Francis recorded is often extremely brief. Nonetheless, Serjeantson's careful examination revealed that during almost the entire time Willughby was at Cambridge, he returned to his commonplace book again and again to add notes on what he had read. It is striking how closely Francis followed Duport's advice: 'Transcribe not whole sayings & stories at length, for that is tedious and Endless; but make short references to Book and Page.'[25]

Duport further suggested that when studying a particular book, students should read it through in its entirety and in doing so 'observe ye most remarkable passages and note them with a black lead pen [pencil] and afterward refer to them in y^e commonplace booke'. That Willughby followed this advice is clear from the annotations he later made in his copy of the *Theatre of Insects*, whose author, Thomas Muffet, we shall meet

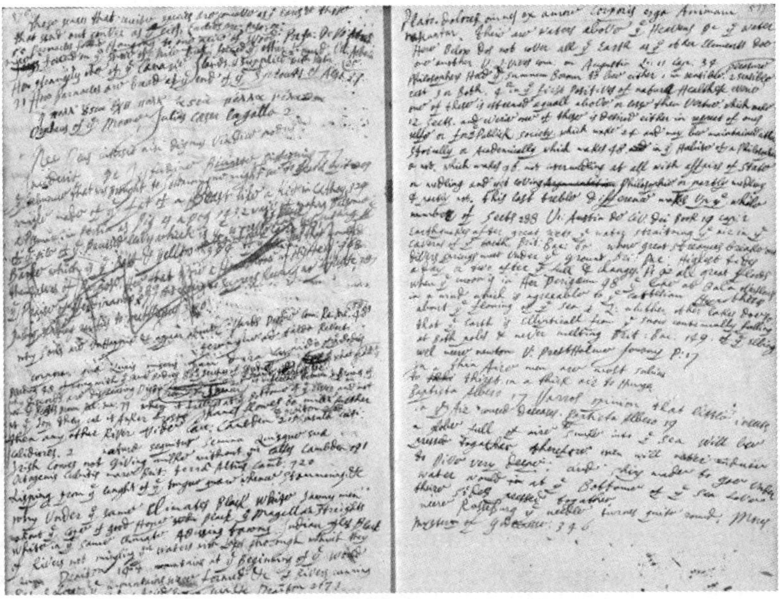

Pages from Francis Willughby's commonplace book.

later.[26] In this instance, Francis's comments are brief, or comprise simple marginal marks that highlight issues of particular interest, and contrary to Duport's advice they are in ink rather than pencil.

A decade after Willughby left Trinity, James Duport wrote a poem about him, as he did for many of his friends and tutees, in which he anticipated how passionately Francis would pursue his studies:

I foresaw this long ago at Trinity College
(where you too, as Noblemen are not accustomed to do,
Eagerly taking both degrees, have adorned the Gown for a very long time),
What a glutton for books and culture
You would be: for there I remember well
That for you, panting heavily for education,

Hastening to acquire skills with too swift a foot,
There was a need for reins, not for spurs.[27]

It was true. When Francis graduated with a Bachelor of Arts degree in January 1656, he was top of his year and ranked above every other student at the university. Exactly what this means is slightly unclear, because the order in which graduating students were listed was determined largely by their social standing rather than their academic prowess. Yet in Willughby's case it seems his position at the head of the list reflected both.[28]

As was becoming increasingly obvious, Willughby had a taste for study, and after acquiring his BA – itself unusual among Fellow-commoners – he remained at Trinity to pursue a Master of Arts, which he completed with similar facility some three years later.[29] Judging from the entries in his commonplace book, gaining his BA marked a transition in Francis's reading. Rather abruptly, around 1655–6, he began a course of study focusing largely on the new science, reading Galileo, Henry More, Thomas Hobbes, William Gilbert, more of Francis Bacon, and of course Descartes. This may not simply have been due to the realisation that he enjoyed and excelled in academic endeavour. Change was in the air at Trinity, where the gradual shift from old to new philosophy was gathering pace and would very soon supplant it. This was 'the most radical change in theory and the most fruitful in practice'.[30]

Written like this it sounds much less exciting than it undoubt-edly was. However, I can vouch for how exhilarating this must have been, for I have lived through a similar change in my own area of science. I will try to explain. Following, rather distantly, along the path first cleared by Willughby and Ray through the tangled complexities of natural history, in the early 1800s Charles Darwin came up with the idea of 'transmutation' or evo-lution, mediated through the process of natural selection. The idea, presented in his book *On the Origin of Species*, published in 1859, changed the way we view the natural world, and Darwinian evolutionary thinking has been the bedrock of almost all natu-ral history research ever since. But exactly how natural selection worked remained unclear. Then, in the mid-1960s, after a century of new knowledge, the mechanism started to become apparent. Natural selection operated on individual organisms rather than on populations or entire species as was once thought. The result was the biggest change in twentieth-century biology. During the early 1970s zoologist Richard Dawkins at Oxford wrote *The Selfish Gene* to make this new thinking on 'individual selection' – which he caricatured as 'selfish' – accessible to a wide audience. I was fortunate enough to be a D.Phil. student in Oxford during this time, and attended seminars in which Dawkins tried out his still unpublished chapters and introduced us to these new ideas. There was a strong sense that something extraordinary was happening in biology – as if a bright new horizon were opening. The new science of 'individual selection' created a new world of opportunities that allowed me, and other young researchers, to make sense of those parts of the natural world we were study-ing (birds in my case), which had previously seemed obscure. Willughby and his colleagues must have felt exactly like this too in the 1650s as their new science began to emerge.[31]

Predictably, as with any major change in thinking, the old guard in Willughby's day was sceptical and suspicious; and too set in their ways, perhaps, to learn new tricks. But that scepticism was impor-tant for refining the ideas, for sharpening thinking and ensuring that the new science did not go off half-cock.[32] Duport, almost thirty years older than Francis, was certainly wary of the new sci-ence, and good-humouredly chided him for what he considered his infatuation with Descartes, astronomy and botany:

> Let not the new Philosophers, my Friend, please you too
> much.[33]

In truth, Duport was not hostile to these new ideas, but he was con-cerned that they might eclipse 'more polite learning' and religion.[34] There is even a suggestion that Duport may have been secretly persuaded by his tutees, Francis Willughby, John Ray and Isaac Barrow, that the new science was important, for after his death,

Duport's library was found to contain a number of scientific books, all of which were either purchased or published after the new science had become established.[35]

Duport's 'chiding' of Willughby may have been motivated as much by his concern over his tutee's health as his fear of a new thinking. Bitten by the snake of learning, Francis threw himself into his studies, but Duport, seeing his tutee's unquenchable thirst for the new science, was concerned for his well-being. Duport's long career as a caring tutor must have allowed him to recognise both Willughby's delicate constitution and the dangers that such single-mindedness posed. Later he wrote another poem for Francis expressing his fear:

> Desist a little, grant a pause to knowledge;
> How long will you torment yourself with Empirical Studies
> And with Botanical skill, yielding the field to no one?
>
> Be mindful, I entreat you, also of the Ancient Sage,
> *Nothing in excess*: certainly there is a limit to knowledge too;
> The dropsy of learning is a kind of intemperance.
> Lay down your bow, which if it is stretched too much
> Will immediately be broken: Look after your little body,
> That which is certainly not made of iron or of steel,
> Nor mighty in strength, but quite feeble:
> Have consideration for yourself; take care of and cherish your skin.[36]

There is no evidence that Willughby heeded Duport's advice regarding his health, but he did seek friendship among those with similar values, and this meant that his friends were likewise infected by the new science. One of those he became particularly close to was his cousin, Peter Courthope.[37] The two travelled together – for unknown reasons – sometime in 1658, and clearly got on well. So much so that Francis later entreated Courthope to join him on his subsequent travels, but without success.

Philip Skippon arrived at Trinity in 1655 and despite his family's rather different political background – Skippon's father, Sir Philip

Skippon, had been Cromwell's major-general during the Civil War – Francis and he quickly became friends. Skippon also became friends with Courthope.

Another ally was Nathaniel Bacon, a Fellow-commoner at St Catherine's College, Cambridge. Described as a 'resty fellow', Bacon was a lazy student who 'broke out into some extravagancies', causing his father to take him home. Bacon was rescued by his Trinity tutor, John Ray, who Bacon's father employed to provide private tuition, with the result that within a few years Nathaniel was sufficiently improved that he was to spend many months as Willughby's travelling companion.[38]

Of all those he met at Trinity, Willughby's greatest friend was John Ray. They were to work together for much of the next decade or so in what was to become one of the great partnerships in biology. Theirs was an extraordinary collaboration and, as we will see in the following chapter, neither of them would have been as productive without the other.

2

John Ray and the Cunning Craftsmanship of Nature

More likenesses of John Ray exist than of Francis Willughby, as is fitting for someone who over five decades of research and writing helped – with Willughby – to change the face of natural history. Yet only two of these portraits were done from life. Both show Ray as a rather sick man in his sixties, making him look much older than his age. What's more, our perception of the relationship between Ray and Willughby feels to me distorted by their respective portraits: Ray in his sixties; Willughby, as we will see, in his twenties, exaggerating the eight-year difference in their true ages and reinforcing the widespread perception of Ray's intellectual seniority. The better of Ray's two portraits, executed in coloured chalk by William Faithorne around 1690, shows Ray to be slightly built, with a long, downward-sloping nose, a wispy grey moustache, grey shoulder-length hair and olive-brown eyes. He stares, somewhat quizzically, at the viewer, such that I see a serious-minded, somewhat impatient sitter, and someone I am unable to envision as a young man.

Faithorne's somewhat impressionistic portrait was almost certainly commissioned so that engravings could be made from it, and Ray must have judged it a reasonable likeness, for it is the one that serves – as an engraving – as the frontispiece to several of his books. Yet I found the difference between the smudgy colour image and the hard-lined engraved images disconcerting. With only the chalk image to go on, the engravers seem to have enjoyed some artistic freedom, elaborating where elaboration was probably not needed. Subsequent engravers made renderings of previous engravings in a visual version of Chinese whispers, such that the portraits become less and less of a likeness.[1]

John Ray – spelt Wray until 1670, when he dropped the 'W' to facilitate Latinising his name – was Francis Willughby's most important intellectual ally at Trinity College. They met for the first time in 1653, two terms after Willughby first came up to Cambridge. Ray was to become such a significant feature in Francis's life that it

John Ray, an engraving by William Elder (1694) based reasonably closely on Faithorne's coloured chalk portrait (c. 1690).

is impossible to think about the latter without also knowing about the former.

Ray's father, Roger, was a blacksmith – a craftsman and crucial member of a community dependent on horsepower. John's mother, Elizabeth, was a 'herb-woman' – a collector of medicinal plants. It was from her that he acquired his love for plants, and from his father, a fascination for how things were constructed. John Ray was born on 29 November 1627 at Black Notley in Essex and, presumably because his parents understood the value of a good education, he was later sent to Braintree Grammar School. There, recognising his talent and aptitude, Samuel Collins, Vicar of Braintree, encouraged Ray in his studies and arranged for him to go up to Cambridge at the age of sixteen on 12 May 1644. For unknown reasons, however, the antici-pated position at Trinity College never materialised. Undaunted, the ever-vigilant Collins found a bequest that provided Ray with a scholarship at St Catharine's College, Cambridge, which he entered a month later. After two years, and realising that St Catharine's had failed to provide Ray with the intellectual stimulation he needed, Collins intervened once again and arranged for him to transfer – suc-cessfully this time – to Trinity College on 21 November 1646, where, as sizar, he became the tutee of James Duport.

It is perhaps not surprising that Ray should become friends with another student entering Trinity that same year, the sixteen-year-old Isaac Barrow. Previously, at Charterhouse School, Barrow had distinguished himself only as the school ruffian, despite the fact that his father had paid the headmaster Robert Brooke double fees to keep his son in check. Barrow senior's utter despair at his son's pugilistic tendencies is clear from the fact that he 'often solemnly wished, that if it pleased God to take away any of his children it might be his son Isaac'.[2] Isaac Barrow entered Trinity as a sub-sizar, also under the tutorship of James Duport, and he and Ray may have shared rooms. Duport later described Ray and Barrow as 'the two most brilliant pupils of his whole career'.[3]

Ray obtained his BA in 1647/8, and a minor fellowship the next year, followed by a succession of Trinity College posts including Lecturer in Greek (1651), Mathematics (1653) and Humanities (1655), becoming college steward in 1659. During his time as mathematics lecturer Ray arranged for Francis Willughby to be tutored by Barrow. Mathematics was a central part of the Cambridge curriculum, and two better teachers would have been hard to find. It is no coincidence that as part of his wholesale acceptance of the new science in which mathematics played such a pivotal role, Francis should strive – as is clear from the extensive if almost undecipherable notes in his commonplace book – to become an effective mathematician. Willughby and Barrow formed a close relationship; they regularly ate together at Trinity, and when in 1655 Barrow published his edition of Euclid's *Elements*, he dedicated it to Willughby, together with two other students, Edward Cecil and John Knatchbull. The numerical skills that Francis acquired as a result were later reflected in his studies of games – in particular the 'doctrine of chances' in dice and cards – and in his bold attempt to solve a high-profile mathematical problem on motion.[4]

Seeing 'purposefulness' in nature had been widespread even before the first century AD, but it gained increasing popularity during the medieval period.[5] Once Francis Bacon set the scientific revolution in motion in the early 1600s, this so-called natural theology was forced to change because of the ways in which scholars now acquired information. Logic, direct observation and experiment not only challenged the established 'knowledge' of ancient authors, it also challenged the existence of God. The magnitude of this change is reflected in the fact that by 1660 no fewer than ten books on natural theology had been published, because of the need for many to marry the new science's reason with Christian views. Natural theology, or physico-theology as it was also known, provided that union, using the natural world as evidence both for the existence of God but also for His power and wisdom.

Ray, interested in both the purpose and interpretation of knowledge, embraced natural theology with enthusiasm and read many of the recently published books on the topic. Their titles, including John Wilkins's *Reason and Christianity* (1649), Walter Charleton's

The Darkness of Atheism dispelled by the light of Reason (1652), and his Cambridge colleague Henry More's *An Antidote against Atheism* (1653), all reflect the widespread concern that the new science was fostering atheism. Ray was galvanised by More's proposal that the sophisticated design so apparent in all animals and plants, together with the way their design fitted them to a particular environment, constituted evidence for the existence of God. So taken was Ray with these ideas in the 1650s that he used them as the basis for some sermons, or 'morning divinity exercises', that he delivered at Trinity College chapel in the 1650s.[6]

Ray may also have been influenced by a less well-known book by the Flemish Jesuit, Leonardus Lessius, *The providence of God and the immortality of the soul.* Published first in 1631 and reprinted in 1651, Lessius's book sought evidence for the existence of God in the motion of the heavens, the beauty of natural phenomena and – like Henry More – in the purposefulness of animal and plant design. Of insects, for example, Lessius wrote: 'All parts or members in them are wonderfully faire, all most perfectly agreeing and fitting to the functions, for which they were made.' The inescapable conclusion for the perfect fit or design was that there is 'a most wyse and divine Providence'.[7] The English translation of Lessius's book, with the unlikely title of *Raleigh his Ghost*, reveals it to be a polemic against atheism and those allowing religious dissent in the name of civic peace, while at the same time providing evidence for the existence of God. The evidence takes two forms: God's punishment of sinners, and more positively, the link between animal form and function.[8] It is not known for certain whether Ray read Lessius, but the book was in Trinity's library, so it seems possible that he did; he may have been wary of advertising the fact given Lessius's Catholicism and the existing animosity between Protestants and Catholics.

Another thing that prompted Ray to find a way of combining divinity and natural philosophy was the mechanistic philosophy of René Descartes. A key player in the scientific revolution, his main thesis was that ours was a mechanistic universe in which all phenomena could be explained by physical laws. He saw animals as machines, mere automata, and therefore – in contrast to man – incapable of consciousness, physical sensation or emotions. His argument was partly designed to place humans above and distinct from the rest of the animal kingdom, but for someone like Ray with considerable first-hand experience of the natural world, this simply wouldn't do. Later, in a powerful and poignant attack on Descartes's mechanistic ideas, Ray was to ask rhetorically, 'Why, if they are incapable of experiencing pain, do dogs cry out during vivisection?'[9]

Francis Willughby must have heard Ray's sermons on design and the two of them may well have discussed those ideas. Indeed, I can imagine such conversations adding fuel to Willughby's germinating interest in the natural world. Essentially, Ray's physico-theology focused on the purposefulness of nature and why things are the way they are. His position is nicely encapsulated by a phrase he used when he later published his ideas: 'that the material works of God are wisely contrived and adapted to ends both general and particular'.[10] Why, for example, does a swan have a long neck, short legs and webbed feet? The answer, Ray tells us, is so that it can feed on submerged plants while swimming on the surface of the water: God's wonderful design.

Explaining the natural world as the result of God's wisdom has – as this example shows – the potential of being facile and teleological, but this was not the way Ray generally argued his case. In asking, for example, why different bird species breed at different times, Ray suggests that this ensures their young are reared and fledge when food *for that particular species* (my emphasis) is most abundant, again thanks to God's wisdom. In this instance – and in contrast to his predecessors such as More or Lessius – his answer contains some genuine biological insight and, as studies in the twentieth century eventually showed, is basically correct.[11] The fact that Ray was right about the pattern (if not the process – God's wisdom) could be viewed as an example of what historians of science refer to as 'Whiggism' – distorting history through the use of hindsight to selectively identify discoveries important for modern science.[12] In fact, John Ray's observations so often anticipated later findings that there is little risk of our being Whiggish in our assessment of him.

I have a lot of sympathy with Ray and his colleagues who saw God's wisdom as an explanation for the natural world. Today's scientists, however, have little time for Ray's physico-theology and the idea of a God-inspired world, but I can appreciate just how seductive this notion was to the seventeenth century's new philosophers: God's wisdom seemed to make sense and explain so much. But it didn't explain everything, as Ray himself acknowledged, and indeed, long after he and Willughby were dead, when the notion of natural selection came along in the nineteenth century, it replaced the idea of a God-directed universe precisely because it explained so much more. This is the way science proceeds: old ideas are replaced by better ones, based on evidence. In the minds of some, physico-theology and the centrality of 'God' lingers on in the form of 'intelligent design', a position based on conviction rather than evidence and logic.

Ray's preoccupation with science began with botany, no doubt inspired by his mother's interest in medicinal plants when he was a child. Sometime in the early 1650s, in his late twenties, Ray had been ill, both physically and mentally – stressed and frightened by the difficult political situation – and was forced to take a break from his studies. Part of his recuperation included walking and riding, pastimes that provided him with the 'leisure to contemplate ... what lay constantly before the eyes and were so often trodden thoughtlessly under foot, the various beauty of plants, the cunning craftsmanship of nature'.[13] He eventually decided to catalogue all the plants of the Cambridge countryside – to create a local flora. This was far more ambitious than it now seems, for the existing books were all but useless and Ray had little choice but to start from scratch and perfect his own identification skills. He collected specimens, and either dried them as reference material or kept them alive, cultivating them in his tiny Trinity garden. Ray was also keen to encourage friends to share his pursuits,[14] and three Cambridge colleagues – John Nidd, Peter Courthope and Francis Willughby – all helped, most probably by finding plants for him. It was inevitable that while walking or riding together they discussed broader botanical issues such as the validity of the doctrine of signatures; whether all plants reproduce by seed; the meaning of tree rings; and most significantly of all, what constitutes a species.[15]

The doctrine of signatures was an idea dating back to the days of Pliny the Elder and developed in the sixteenth century by the Swiss-German physician Paracelsus, suggesting that the growth form of a plant comprises a sign or signature regarding its curative properties. Lungwort, for example, with its pale-spotted leaves, resembles a human lung and hence signals its suitability as a cure for lung disease. Similarly, the roots of certain orchids resemble human genitals and thus provide an obvious cure for sexual problems. Much of it was nonsense of course, but in Willughby and Ray's day, the belief that God – the creator and healer – had placed those plants on earth for a purpose made a great deal of sense. And of course, as is now apparent, many of those remedies really worked.

In total, Ray collected and identified no fewer than 558 Cambridgeshire plants, publishing his results in 1660 as the *Cambridge Catalogue*. In the Preface he wrote: 'We would urge men of University standing to spare a brief interval from other pursuits for the study of nature and of the vast library of creation'. He continues: 'Surely we can admit that even if, as things are, such studies do not greatly conduce to wealth or human favour, there is for a free man no occupation more worthy and delightful than to contemplate the beauteous works of nature and honour the infinite wisdom and goodness of God.' He adds: 'Of course there are people entirely indifferent to the sight of flowers … or if not indifferent at least pre-occupied elsewhere … they devote themselves to ball-games, to drinking, gambling, money-making, popularity-hunting. For these our subject is meaningless.' He continues:

> We offer a hundred banquets to the Pythagoreans or rather the true philosophers whose concern is not so much to know what authors think as to gaze with their own eyes on the nature of things and to listen with their own ears to her voice; who prefer quality to quantity, and usefulness to pretension.[16]

It is clear that Ray's engagement with the natural world was not solely academic; it was also spiritual, emotional and aesthetic:

> First I was fascinated and then absorbed by the rich spectacle of the meadows in spring-time; then I was filled with wonder and delight by the marvellous shape, colour and structure of the individual plants. While my eyes feasted on these sights, my mind too was stimulated. I became inspired with a passion for Botany, and I conceived a burning desire to become proficient in that study, from which I promised myself much innocent pleasure to soothe my solitude.[17]

I can hear Ray uttering these very words to Willughby as they botanised together, and it isn't difficult to see how Ray's passion for natural history ignited Willughby's. Their joint excursions must have brought the two men closer together; the younger Willughby in awe perhaps of Ray's knowledge and clarity of thought; the older man in awe of Willughby's social status, but probably even more, his receptivity, aptitude and sharpness of wit.

As well as botany, other practical activities helped to shape Francis's intellectual development. These included anatomy, inspired by the works of William Harvey on the circulation of the blood, and on the reproduction and development of animals; and by Descartes's 'mechanistic' view of nature. The new scientists at Trinity College – John Ray, Isaac Barrow, Walter Needham and John Nidd – were intrigued by the internal structure of animals, and on one occasion at least and probably with Willughby present too, they dissected several birds including a bittern, a curlew and a 'yarwhelp'.[18] Ray commented that 'The yarwhelp is a name I never read or heard of before or since.' It was in fact the East Anglian name for the bar-tailed godwit and clearly distinguished from the similar black-tailed godwit, which was then referred to simply as the godwit.[19] It seems likely that the group also anatomised other animals (but not humans) and that it was at Trinity where Francis Willughby and John Ray acquired their fascination and aptitude for dissection.

The idea of John Ray, Francis Willughby and friends dissecting a bittern in John Nidd's rooms caught my imagination. I have dissected many birds myself, and I began to fantasise about what that must have been like and decided to relive the occasion by dissecting my own bittern. Easier said than done. The bittern is not only a rare bird in Britain, it is also extremely secretive and more often heard than seen. Its booming call was well known in Willughby's day, as the reedbeds on which the species depends were more extensive then. Drainage of the fens and wetlands subsequently drove the bittern to extinction as a British breeding bird in the late 1800s. Since then it has become re-established and now, through careful conservation, the population in Britain numbers some 150 booming males. The bittern population is counted as the number of calling males rather than as pairs because the species is both polygamous and impossible to see and count inside their *Phragmites* reedbeds.

Dead birds are often handed in to museums, so I enquired whether any museum in the United Kingdom had a bittern in their freezer, but none had. I next asked the managers of reserves where bitterns breed if they had a dead one: none did. But then, to my amazement, a few weeks later a reserve manager called me to say he had just found a bittern that had died after colliding with some overhead wires.

My first reaction on seeing the bird was how small it seemed. Small, but magnificent. Having watched bitterns magnified through binoculars or a telescope, one's brain is duped into expecting a larger bird. This is a bird with golden, star-spangled plumage; a dagger-like bill; leaf-green skin on the upper eyelid and gunmetal blue beneath. Its long, lime-green legs are tipped with elongated toes terminating in extraordinarily long claws, one of which is exquisitely serrated. All these features are adaptations for a life among the reeds; the bird's plumage provides protection through camouflage; its beak and claws enable the bittern to spear and hold the slithering eels on which it feeds; and those long legs, toes and claws allow the bittern to clamber confidently through its thatch-like reedy habitat. The serrated middle claw is thought to help remove eel slime from the bittern's plumage.

The bittern's physical features noted by Willughby eventu-ally appeared in the *Ornithology*. As well as the bird examined in Nidd's rooms, he and Ray scrutinised others, including one whose irises were 'hazel incline[d] to yellow' and another with red eyes. Willughby also comments on the bittern's large ear opening, which I also made a point of looking at by parting the feathers behind the eye. The bittern's ear is unusual in that it is an oval of bare, blue-purple skin with a relatively small opening, suggesting to me that there might be something special about the bittern's hearing – pos-sibly linked with its exceptionally loud call.[20]

The bittern's booming has been likened to 'the lowings of an ox'[21] and accounts for the first part of its scientific name *Botaurus stel-laris*, from 'bos', meaning ox, and 'taurus', meaning bull. The name *stellaris* refers to the starred or mottled plumage. Willughby calls it the 'Bittour or bittern or mire-drum'. 'Bittour' is Old English, and Chaucer in the *Wife of Bath's Tale* mentions that 'a Bittour bumbleth in the mire'.

I haven't quite finished with the bittern's external appearance. Parting the buff and streaky breast feathers, I expose two well-concealed patches that at first sight look like shaggy, white fabric, covering much of the bird's breast. These patches are so striking and so unlike anything I've ever seen on any bird before, I cannot quite believe that neither Willughby nor Ray commented on them. They *are* well concealed, but knowing how meticulous Willughby was with his examinations and descriptions, I'm still surprised that he overlooked them or missed reporting them. These are patches of highly modified feathers known as powder-down, designed to dis-integrate almost as soon as they emerge from the skin, generating a coarse talcum-like dust thought to aid the waterproofing of the feathers. It probably also helps to coat the copious slime from the eels on which bitterns feed, so that it can more easily be removed from the plumage.

Inside the bittern I discover an out-of-season ovary (the bird died in early December), attached to a surprisingly large oviduct, indicating that this was a mature female. Willughby describes the windpipe or trachea with its incomplete cartilage rings; the two-lobed liver, the stomach 'of a singular structure, and of the figure of the letter S', but, curiously, he makes no comment on the gut itself. I knew from an earlier account that the bittern has a long alimentary canal. 'Five ells long' is how that previous writer describes it. An ell was a German unit of measurement, generally meaning the distance from a man's elbow to his fingertips, or about forty centimetres. So five ells is two metres, and sure enough when I carefully freed my bittern's gut from its connective tissue, and stretched it out, it was indeed two metres in length. Once again, I was surprised that Willughby had not noticed this. The nineteenth-century ornithologist William MacGillivray used – very much in the Willughby tradition – anatomical features to classify birds and recognised that a long, narrow gut was a characteristic feature of the heron family, to which the bittern belongs.[22]

Among the various extra-curricular activities Willughby and Ray engaged in at Trinity College was 'chymistry', a field of investigation closely allied with medicine, and whose unfamiliar name reflects its slightly uncertain position between alchemy and chemistry. The study of chymistry and the use of chemicals in the preparation of medical remedies had been pioneered by Paracelsus. Although little known, chymistry appears to have been popular among certain of the new scientists at Trinity, including Willughby and Ray. Others involved were Alexander Akehurst, the vice-master of Trinity (who in 1654 was ejected for blasphemous statements); Daniel Foote, admitted as a sizar in 1655, the year before Willughby; Francis Jessop (admitted in 1654), whose wide range of interests included ornithology; Thomas Pockley, another of Willughby's contemporaries, who was, in addition to being a chymistry enthusiast, passionate about anatomy; and John Nidd, another anatomy enthusiast whose chymistry books were much in demand by his friends.[23] After Willughby and the others had left Cambridge, Trinity's tradition of ex-curricular chymistry continued with Isaac Newton in the 1660s.

To foster their students' chymical expertise Trinity hosted scholars from abroad. These included the Greek, Constantine

Rhodocanacis, who later patented a general remedy called 'true spirit of salt', which is hydrochloric acid and was thought to be useful, therefore, in restoring the balance between acid and alkali humours.

As this suggests, chymistry involved the production of medicines, but it also involved the transmutation of elements, and it seems that Francis Willughby and John Ray were involved in both. Medical tinctures, which were taken orally, were made by bruising or crushing plants and dissolving their extracts in alcohol. An additional method, referred to as palingenesis, involved bruising and burning a plant and 'calcinating' the ashes to 'reveal' a volatile salt, which when heated, grew – uncannily – in a ghost-like three-dimensional replica of the plant that had just been burned. The salt that emerged was assumed to contain the 'essential virtue' of the plant, which could then be used as a remedy.

In January 1659 John Ray wrote to Peter Courthope to say that he and Thomas Pockley were performing all the 'easie & useful chymicall experiments wch wee find in bookes'.[24] Willughby was also involved, for his commonplace book contains numerous references to experiments under the general heading 'Praeparationes et Experimenta Chimica'. Some of these, as he knew, were blatant nonsense, like mixing 'Two partes of the fat of a goose mingled with one of the salt of a goose' as an 'arcanum [an elixir] to make hair growe' – hardly one that Willughby needed anyway, judging from his portrait. Of the more 'serious' remedies, one included the making of 'salt of tartar', which was variously used as an aperitive ('ten to thirty grains' was the dose),[25] a cosmetic or as a treatment for cancer.

A heading in Willughby's commonplace book, 'Mr Wray's Experiments', included tests with antimony, which in various forms such as antimony cups, drinking glasses and tinctures, had been used as an emetic since at least the first century AD as a way of restoring the supposed balance between the humours. After the publication in 1604 of Basil Valentine's book, *The Triumphal Chariot of Antimony*, the popularity of antimony increased still further.

The idea of humours was ancient, dating back to the fifth century BC when Hippocrates decided that our four humours – blood, yellow bile, black bile and phlegm – provided a framework for understanding human health. Each was linked, not only with heat, cold, dryness and moistness, but also with the elements: air (blood), fire (yellow bile), earth (black bile) and water (phlegm). We shouldn't be too surprised by the Hippocratics' obsession with bodily fluids since sick individuals typically exhibit one or more of the following symptoms: a runny nose; sweating; a change in urine colour; or coughing up phlegm or blood – clear evidence that the humours were out of alignment. Physicians sought therapies that would rebalance the humours, by purging, vomiting, sweating or bleeding.[26]

On 28 November 1658 Willughby noted his success in creating metallic antimony, known as 'regulus of antimony', that is, 'the little king', and so-called because antimony combines readily with the king of metals, gold. Creating regulus of antimony was difficult and Willughby and Ray made several attempts with different combinations of chemicals before succeeding. Its production was important because, through the doctrine of signatures, its star-like structure suggested that antimony's enduring medicinal powers were drawn directly from the enduring light emanating from the celestial stars. Regulus was a vital step in creating the elusive philosopher's stone, knowledge of which God had first given to Adam. If acquired, the philosopher's stone would allow one to transform base metals into gold, to become immortal, and also to create a universal medicine. Since the transformation of any metal into gold was strictly illegal in the 1650s and 1660s, I like to think that Willughby and Ray's quest was for a universal medicine.

Rhodocancis was experimenting with different tinctures of antimony at this time. In a letter to Willughby dated 17 March 1660, Peter Courthope wrote: 'Constantinus Rodocanasis [sic] salutes you ... This morning he was in my chamber, and gave me a taste of a Tincture of Antimony, which had no acrimony at all, yet deep and strong, which he prizes very much: He only told me it was prepared of the Glass of Antim[ony]. But how I know not yet.'

In the same letter Courthope went on:

> Yesterday he brought his materials into the Combination [dessert], and shew'd the Experiment of the Tree, which suddenly arose, and within less than an hour reached the top of the liquor: [It] especially was of a quick growth, which sent forth a long shoot above the liquor, which was much bigger at the top than the bottom, that the weight of it brake it off.[27]

The 'Experiment of the Tree' was the brainchild of the German alchemist Johann Rudolf Glauber. He is better remembered for having discovered sodium sulphate in Austrian spring water, which he called *sal mirabilis* – miraculous salt – because of its wonderful laxative properties and was later named Glauber's salt. Glauber had discovered the Experiment of the Tree – also known as the silica garden – in the 1640s. This is a remarkable phenomenon and one I still have a vivid memory of from my early schooldays. It consists of placing crystals of metal salts such copper sulphate, cobalt chloride or iron sulphate into a solution of potassium or sodium silicate (that is, waterglass). Within minutes the crystals start to grow into remarkable plant-like forms, an effect enhanced by the fact that different salts produce trees of different colours: copper sulphate, blue; cobalt chloride, purple; iron sulphate, orange. Glauber described the forms that slowly surged and branched upwards through the clear viscous fluid as 'metallic trees' and 'mineral vegetation'. The chemical changes that result in this rapid tree-like growth are now understood.[28] In Willughby's time it was a spectacle of beauty and wonder, but significantly, also a link between botany and chymistry indicative of some kind of 'life force' within metals. It was the possibility of a life force within minerals that – as we shall see – later fuelled confusion in the debate about the origin of fossils.

The flourishing of the new science at Trinity during Willughby's undergraduate days had its origins in Oxford – the 'other' university – in the late 1640s, when 'Several diverse ingenious persons ...

used to meet at the lodgings of the excellent person, and zealous promoter of learning, the late Bishop of Chester, Dr Wilkins.'[29] Inspired by the teachings of Francis Bacon, Wilkins – then warden of Wadham College, Oxford – encouraged a group of like-minded individuals to perform experiments to better understand the natural world. These were exciting times for philosophy, but following the execution of Charles I in 1649, difficult and complicated times politically. A decade later, around the time the monarchy was being restored, key members of the group reformed in London at Gresham College. And it was here, on 28 November 1660, following a lecture by Christopher Wren, that twelve of them – with John Wilkins as chairman – agreed to form 'an association for the promoting of experimental philosophy'. Among those present were Wren, Robert Boyle, Mr Bruce, Lord Brouncker, Sir Robert Moray, Sir Paul Neile, Dr Goddard, Dr Petty, Mr Ball(e), Mr Rooke and Mr Hil(l).[30]

Proposed by John Wilkins and John Ray, Francis Willughby was admitted in December 1661 and formally elected as a Fellow of the Society on 20 May 1663: he was twenty-seven. Willughby is listed in the Society's records as an 'Original Fellow', that is, one of those elected before the Society received its Royal Charter later that year, when it became 'The Royal Society of London for Improving Natural Knowledge'. The Society was to play an essential role in Willughby's development as a scientist.

╞ᛝ╞

What was Francis Willughby like? We don't have a great deal to go on, mainly a portrait and a eulogy, but enough, I think, to create a sense of who he was and how his friends saw him. Physically, we have an undated portrait attributed to Gerard Soest, a prolific and popular portrait painter of seventeenth-century minor gentry. Judging from Willughby's appearance, the portrait could have been completed at any time between his twenties and his death at thirty-six. However, since Soest was based in London it seems likely that the portrait was painted when Willughby was also there – on and off – between 1657 and 1660, that is, aged between twenty-two and twenty-five. The image, in which Francis is holding a book and marking a page with his index finger, also implies that the painting was made after he had decided on a life of study. With long, fair locks – which were then fashionable – and a short fringe, he is clean-shaven, with dark eyes and a handsome nose. In terms of his expression, it is a pleasant portrait, but not one of Soest's best (nor worst), for the artist has given him a slightly vacant countenance, and overall, as one might expect, Francis appears somewhat serious; but then he was more than somewhat serious about his work. The portrait is hardly a Velázquez, sadly. Had it been, and given the dates it could have been (just), we would have a psychological portrait in paint. Through the proficiency of his penetrating analysis, palette and brushes, Diego Velázquez would have given us the painterly equivalent of a photograph snapped at the instant that Willughby's personality, aspirations and motivation were apparent. Instead, we have Soest's mere likeness. Others, more imaginative, have seen more in it than I can, and viewing this portrait in the early 1800s, the ornithologist William Jardine, a Willughby devotee, declared it the 'beau ideal of a naturalist's countenance'.[31]

This aspect of Willughby's personality is very clear from John Ray's generous memoir in the preface to their ornithology volume, penned just a year or two after Francis's death. Ray records a heart-warming set of attributes, starting with the fact that, notwithstanding Willughby's inherited social position, he strove through his studies to create his own modest identity. As Ray says, 'God had given him quick apprehension, piercing wit and sound judgement', which by 'his great industry he did highly improve and advance'. Willughby's industry defines him to a large extent, for he was 'since childhood addicted to study' and 'detested no vice more than idleness'. James Duport recognised Willughby's energetic, goal-orientated approach to life. So did John Wilkins, who when staying with Willughby at Middleton wrote to a colleague saying that Francis was 'as much in love with study & experiments, as ever any man was with a mistress'.[32] As is now apparent, once Francis

had discovered natural history his enthusiasm and diligence knew
no bounds. He took to heart the advice Duport gave to his tutees
about living each day as though it was their last. The result was,
as Ray says, that no one else he knew was as knowledgeable about
'birds, beasts, fishes and insects'.[33]

Somewhat remarkably for someone so obviously driven, he
seems also to have been extraordinarily humble. Ray tells us that
Francis was:

> ... endowed with excellent gifts and abilities both of body and
> mind, and blessed with a fair estate. Howbeit, as he did duly prize
> these advantages of birth, estate, and parts, so did he not content
> himself therewith, or value himself thereby, but laboured after
> what might render him more deservedly honourable, and more
> truly be called his own, as being obtained by the concurrence at
> least of his endeavours.[34]

Francis also respected men of all persuasions, regardless of their
wealth or position; he was sober, temperate and 'never tempted to
excess, scrupulously just, true to his word and loyal to his friends'.[35]

Even allowing for the fact that Ray was honouring his dead
friend and may have exaggerated for the family's benefit, if only
half of what he says is true, Willughby was evidently a nice man. I'd
love to know more. I'd like to think that in addition to his industry
he was sufficiently relaxed, at least at times, to laugh and joke with
his friends. We know from a few comments in Ray's writings that
Ray had a sense of humour and it is difficult to believe that two
men would work so closely together and for so long without enjoying a few laughs.

Willughby died when his daughter Cassandra was just two,
so her description of her father's character in her essay on
the family's history is based almost entirely on an account of
Cambridge alumni and on what Ray said in the preface to the
Ornithology, and sadly, almost nothing directly from her mother.
As though to emphasise this, Cassandra says she wished Ray 'had
been more particular in his account of my father, for doubtless

many remarkable things of his life had been very well worth his notice, and he, by spending severall [sic] years with him, might have noted down to posterity such useful examples of his life, as now I have in vain endeavoured to recover'. The only things she reports via her mother were: her father's regular hours of study; his ability – as a result of studying physic (medicine) – to provide medications to all his neighbours; and his 'compassion and goodness to the poor'. As a sheriff, Willughby was expected to uphold the law and could not officially tolerate beggars or vagabonds. To avoid having to take action against them when they came to the Middleton kitchen in search of food, he simply arranged not to 'see' them.[36]

Francis Willughby's formal Cambridge education was extended and complemented by botanical fieldwork, bird dissection and chymical experiments. As is so often the case, it was these extra-curricular activities with their sense of novelty and risk, and their being part of something ground-breaking, that appealed to Francis Willughby. This rich and stimulating mix, undertaken with the guidance and companionship of John Ray and a cluster of close friends, set Francis on a course he would pursue for the rest of his short life.

3

A Momentous Decision

During Francis Willughby's last few weeks in Cambridge in January 1660, Trinity College appointed the clergyman and natural philosopher John Wilkins as the new Master. Wilkins arrived with an idea that would change the way Willughby thought about the natural world.

Even though his appointment at Cambridge was a gift from Oliver Cromwell, Wilkins was both politically and religiously tolerant and had been instrumental in calming the considerable political tensions prevalent during the interregnum. A lover of mankind, Wilkins was described by a contemporary as 'the wisest of clergymen'.[1] In the portrait that now hangs in Wadham College, Oxford, where he presided as Warden between 1648 and 1659, Wilkins appears pious and reliable, yet the flowing locks, chubby cheeks and steady stare reveal little of the man himself. Intelligent and passionate about the new science, Wilkins was considered more as a facilitator than an innovator. He had, however, a great sense of fairness, and, although married to Cromwell's widowed sister, both Roundheads and Royalists alike respected him. Wilkins's marriage to Robina in 1659 was an improbable union, for she was sixty-two and Wilkins forty-two, but it seemed to be based in part at least on genuine affection, and, crucially, it helped to protect the universities of Oxford and Cambridge from 'the ignorant, sacrilegious commander and soldiers' (Cromwell

and his army) who would have otherwise destroyed them.[2] As a
key player in the founding of the Royal Society, Wilkins ensured
that it accepted Fellows regardless of their Christian views. Robert
Hooke, with whom Wilkins worked closely at the Royal Society,
and who shared his passion for science, summed up Wilkins in
these fine words:

> A man born for the good of mankind, and for the honour of his
> country, in the sweetness of whose behaviour, in the calmness
> of his mind, in the unbounded goodness of his heart, we have
> an evident instance, what the true and primitive unpassionate
> religion was, before it was soured by particular factions.[3]

As someone who thought deeply about science and was excited
by the 'new philosophy', Wilkins also recognised that there was
a problem. It was that there was no universal language of science,
and since the 1630s he had been interested in creating an artificial
language that everyone everywhere would understand; one that
avoided ambiguity and allowed for the ever-expanding lexicon
necessitated by new methods and new discoveries in science.

The notion of a universal language was not new, however. Italian
scholars during the Renaissance had thought about it, and the idea
had also been popular among French savants. When it eventually
reached England in the early 1600s, Francis Bacon quickly became
a devotee. He could see the obvious advantage of a system in which
one word represented one thing, rather than multiple words for a
single thing, or different things known by the same name. Bacon's
motivation was clarification. Not only did the new science need to
be increasingly precise in what was meant by specific terms; new
knowledge, including new species, also needed new words. In their
current state, Bacon recognised, language and terminology were in
a mess. Not only were different terms needed for different things
(such as the different feathers on a bird's body), but some things,
such as insect species, which had no names, needed names. Central
to all this was precision; more precision required more words. For
Bacon, ignorance about the world created linguistic inaccuracy and

this inhibited understanding of the world. If society was to move forward and take advantage of new scientific knowledge, a universal language was an essential first step.

Like Bacon, Wilkins was religiously motivated and believed the need for a universal language to be deeply rooted in the past. Specifically, that man's arrogance – epitomised by the construction of the Tower of Babel – had been punished by God who confounded people by creating separate languages across the world rather than the one spoken by Adam. The consequence was that although people were able to form a mental image of something – for example, a honey-buzzard – in different regions or countries, they used different arbitrary sounds (words) to identify it.

Wilkins's proposed solution to this linguistic confusion was to create a kind of scientific Esperanto by which everyone would know 'things'. His extraordinarily ambitious plan was to develop a philosophical language based on 'real characters', rather like Egyptian hieroglyphs or Chinese characters, to represent specific things or notions. However, for nature to be represented by a system of symbols, it was essential to discover the elements upon which all natural phenomena were based. Wilkins recognised that creating a philosophical language required a model of the natural world, and for seventeenth-century scientists that model was classification – the ordering of nature. The new science needed a new language just as much as a new language needed science.

The key to the new science was the organisation of knowledge. Although the scientific revolution sought to overturn much of Aristotle's thinking, it was, at the same time, based on two fundamental Aristotelian assumptions. First, that there was order in nature. Second, the idea of an organism's 'essence' – what made it *it*.

For Francis Bacon and the philosophers who followed in his footsteps, that order had been imposed by God and was deeply embedded in Christian belief. It was a belief that order lay at the end of the rainbow. If order existed then it could be discovered, and discovering order was the basis of the new science. Discovery, however, depended on an objective way of interrogating the natural world and this meant that science needed a 'method'. In natural history it needed a methodology that allowed one to identify the 'essence' of things, that is, what, for example, made one bird species similar to another, yet distinct from others? Only by knowing such things could one impose order, and for seventeenth-century natural historians such as Willughby and Ray, classification was to become their core activity. For the physical scientists on the other hand, such as Descartes and Newton, mathematics was what enabled them to verify the existence of order in the natural world. Order was uppermost in many people's minds. The Civil War had created monumental and awful disorder, so consciously or unconsciously the quest for order was paramount and classification and quantification became the foundation of the new science.

The restoration of the monarchy in May 1660 brought Wilkins's Cambridge career to an abrupt end when the position gifted to him by Cromwell was retracted. Wilkins eventually bounced back, becoming Bishop of Chester in 1668, the same year that his 'universal language' project – of which more later – was published.

Ray, meanwhile, was still at Trinity, continuing with his botanical studies, and preparing his *Cambridge Catalogue*. Over the previous years he had carefully documented not only the plant species that flourished in the Cambridge countryside, but also the habitats – woods, bogs or meadows – they occupied. The text of Ray's modest book is interspersed with intriguing anecdotes, including the idea that tree rings accurately reflect the age of a tree: 'My opinion is that so long as the tree is alive, it adds a new ring, even if it is a narrow one, every year'; how the shape of a tree is influenced by the prevailing winds; that gastropods (slugs and snails) were hermaphrodites; that rose bedegaur galls, also known as robin's pincushion, contain tiny white maggots that emerge as flies 'like those of winged ants' (the maggots are known now to be the larvae of the gall wasp *Diplolepis rosae*); and, finally, that after pupation the caterpillars of the large white butterfly give rise either to a butterfly or to numerous tiny worms that in turn spin cocoons, from which emerge 'flies, black all over with reddish legs and long antennae, and about the size of a small ant'.[4]

Published anonymously in February 1660, the *Cambridge Catalogue* acknowledged the help that Ray had received in the field from Francis Willughby and his friend Peter Courthope. Given the many exciting discoveries they made together, it is perhaps not surprising that Willughby, for one, became completely enthused by natural history.

This enthusiasm resulted in Willughby and Ray undertaking several journeys together. Their first was in August 1660 when they headed north on horseback to Halifax, through Keighley to Settle and on into the Lake District, then to Ravenglass where they crossed the sea to Ramsey on the Isle of Man. From there they visited the Calf of Man, a tiny island (that now has a bird observatory) lying about a third of a mile off the south-west end of the main island. It was on the Calf that Willughby saw and described 'the puffin of the Isle of Man' taken out of a nest. The bird in question was the Manx shearwater, and the individual Willughby saw was a nestling, whose extraordinary appearance as a ball of copious grey down with a beak cannot but have impressed him.

Today, fluffy shearwater chicks are considered 'cute', but I doubt Willughby thought like that. Nor do we know whether Willughby and Ray spent the night on the Calf (probably not), for had they done so they would surely have commented on the cacophony of calls created by the adult shearwaters returning to the colony after dark to feed their single chick. They would undoubtedly also have collected a specimen of an adult shearwater – which they seem not to have done. Instead, their account of this species subsequently published in the *Ornithology* sounds second-hand. That it is Ray who has written the account is clear, for after telling us that Willughby encountered 'only a young one taken out of the nest', Ray tells us that he saw what must have been dried adult specimens in 'the Repository of the Royal Society and in the cabinet of curiosities owned by John Tradescant'. The Royal Society's 'repository' was its museum of biological specimens and scientific instruments, under the care of Robert Hooke and housed at Arundel House

Manx shearwater Puffinus puffinus*: it was a chick probably at this stage of development that Willughby and Ray saw on the Calf of Man.*

beside the Thames on the Strand in London. Tradescant's collection – so vast it required a 183-page catalogue known as 'The Ark' – was one of the wonders of seventeenth-century London and later became the basis of Oxford's Ashmolean Museum.[5]

The account of the shearwaters on the Calf of Man in the *Ornithology* states that: 'This islet is full of conies [rabbits], which the *Puffins* [shearwaters] coming yearly dislodge, and build in their burrows.' Continuing, Ray tells us that they lay a single egg, 'although it be the common perswasion [belief] that they lay two at a time, of which one is always addle … The old ones early in the morning, at break of day, leave their nests and young, and the island itself, and spend the whole day in fishing in the sea, never returning or once setting foot on the island before evening twilight: so that all day the island is so quiet and still from all the noise as if there were not a bird about.'[6]

Ray further informs us that shearwaters 'feed their young ones wondrous fat' and that 'those [people] intrusted by the Lord of the Island [Lord Darby] draw them out of the cony-holes and in order to keep a count of the numbers they take, cut off one foot' (which, Ray says, gave rise to the fable that they are one-footed). They sell, he says, 'for nine pence the dozen' and the 'Romish [Roman Catholic] Church' allows them to be eaten in Lent because they taste so like fish. Finally, Ray recounts that 'Notwithstanding that they are sold so cheap, yet some years there is thirty pounds made of the young Puffins [shearwaters] taken on the Calf of Man: whence may be gathered what number of birds breed there.' A quick calculation indicates that a total of £30 at nine pence per dozen represents 9,600 shearwater chicks. Presumably this is a minimum, for many nesting burrows must have been inaccessible or undetected. In the late 1700s brown rats arrived on the Calf and the numbers of the eponymous shearwaters collapsed, so that they currently number just a few hundred pairs. The common name Willughby and Ray used for this species accounts for the Manx shearwater's otherwise puzzling scientific name: *Puffinus puffinus*.

The illustration of the Manx shearwater that appears on the final plate of the *Ornithology* is one of the most lifelike images in the entire book. And this is precisely because, in contrast to so many of the others, it was a drawing executed from a live bird. The bird in question was one of two shearwaters kept by Sir Thomas Browne, Willughby and Ray's 'learned and worthy friend' – a gentle, intelligent physician with an interest in birds – based in Norwich. Without saying where he got them from, Browne told them, 'I kept two of them five or six weeks in my house, and they refusing to feed, I caused them to be crammed with fish, till my servant grew weary, and gave them over: and they lived fifteen days without any food.' The original image of a (very healthy-looking) shearwater, executed in oils on heavy paper, and probably painted by Browne's daughter Elizabeth, is now in the British Library in London.[7] The remarkable thing about this shearwater (and its description in the *Ornithology*) is that Ray does not connect it either with the nestling Willughby obtained on the Calf of Man, or those dried specimens at the Royal Society that Ray refers to as the 'puffin of the Isle of Man'. Clear evidence that species identification, be it based on birds of different ages, alive or dead, illustrated or real, could be a considerable challenge.

Willughby and Ray returned to Middleton and Cambridge, respectively, in late July 1660.[8] Their journey together inspired and reaffirmed Willughby's passion for the natural world, for soon after he was home he set off for Oxford where he spent several weeks in the Bodleian Library reading the great works of natural history. From Oxford he wrote to his friend Peter Courthope entreating him to visit, but by this time Courthope had left Cambridge and returned to his family home and was unable to accept Willughby's invitation. It may have been just as well that Courthope couldn't go, for it would probably have distracted Willughby from preparing himself for things to come.[9]

It must have been with great eagerness that Willughby and Ray planned their next journey, this time to Wales and Cornwall. They were to be accompanied by another of Ray's tutees, Philip Skippon, whom Ray was concerned might not get on with Francis because of their two families' very different political backgrounds. In fact, Philip and Francis liked each other and the three of them proved excellent travelling companions. And this was the journey that set the course for Willughby's and Ray's lives.

ﾑﾑﾑ

Leaving Cambridge on 8 May – Ascension Day – 1662, Ray headed, via St Neots, to Middleton to collect Willughby en route. As in their previous travels they made notes on the state of the towns and villages through which they passed, as well as all aspects of natural history they encountered. At Sutton Coldfield, for example, they reporting finding 'in great plenty' moonwort *Botrychium lunaria*, a rather beautiful fern with distinctive circular leaflets.

A few days later they visited Norbury Meer, now Nantwich Lake, to see the 'puits' or black-headed gulls, where they either witnessed for themselves or were told that the birds lay clutches of three or four eggs and how 'at the driving every year, they commonly take above a hundred dozen [1,200] young, which they sell at five shillings the dozen'.[10]

An intriguing illustrated account of the 'harvest' is provided by Robert Plot, who visited the 'pewit poole' at Norbury a few years later. Slightly younger than Willughby, Plot graduated from Magdalen Hall, Oxford, was elected to the Royal Society in 1677 and served as one of the Society's two secretaries in 1682, and in the following year became keeper of the Ashmolean Museum

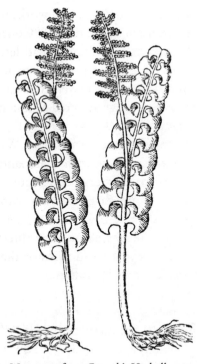

Moonwort from Gerarde's Herball, 1597.

in Oxford. Not renowned for either his scientific accuracy or
his morals, Plot was held in low esteem by Ray, who wrote (in
1692) to a friend saying that Plot 'may be too much influenced by
worldly advantage of honour and profit'.[11] In his *Natural History of
Staffordshire* published in 1686, Plot refers to the 'learned and inde-
fatigable Mr Willughby and Mr Ray', only to draw attention to a
number of birds he says they failed to identify or had overlooked in
the *Ornithology*.[12] It is far from clear, however, that Plot is correct
and it may well have been statements such as this that irritated Ray.

Plot described how the islands on which the puits bred at
Norbury were prepared by cutting the vegetation to provide nesting places and how the birds returned each year to settle at the pool
around Lady Day (the Feast of the Annunciation, 25 March), build
their nests and lay their eggs. As soon as the chicks had reached
sufficient size, but still unable to fly, they were herded into long

rabbit nets and placed in pens where they were subsequently fat-tened on bullock offal or corn and curds. One writer described the birds' lean flesh as 'delicious' with 'a raw gust of the sea', which sounds remarkable! Willughby and Ray's friend, Thomas Browne, felt it rather odd that people should eat young gulls while at the same time refusing to eat 'other animals whose [natural] food was no more impure', but the trick was in the artificial feeding that was said to improve their flavour. It was clear, too, that such habi-tat management, harvesting and artificial feeding had a long his-tory: the Romans did it with thrushes; and Thomas Muffett, author of *Health's Improvement*, said that young gulls 'being fattened … alter their ill nature, and become good'.[13]

The name 'puit' is onomatopoeic and describes the gull's call; the same can be said for the peewit (or lapwing) – although the sounds made by the two species are quite different.

On arriving at Chester, Ray commented that the cathedral church was remarkable for nothing except a preaching place and that the bishop's seat in the choir was made of stone.[14] Willughby and Ray would return to Chester later in 1669, when John Wilkins was bishop there.

From Chester, Willughby and Ray made their way via Wrexham and Holywell to Denbigh, which they considered 'one of the great-est towns of North Wales', reaching Bangor on 19 May 1662, the day Parliament passed the Act of Uniformity. The idea of such an

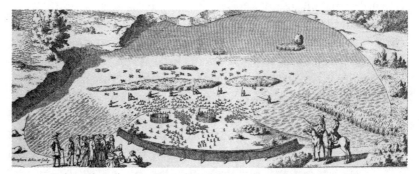

Robert Plot's drawing from 1686 of the puit harvest at Norbury Meer.

Act – designed to quash the non-conformist churches that had pro-liferated under Cromwell – was in the air before they left, but, because they were travelling and news moved relatively slowly, it seems unlikely that Ray and Willughby received confirmation until later. Either way, as will become clear, the Act brought about a major change in Ray's circumstances.

From Beaumaris they took a small boat to Prestholm (Priestholm, known now as Puffin Island in English and Ynys Seiriol in Welsh) on 22 May, noting the plants, including the edible sea-beet *Beta maritima*, and the birds: 'two sorts of seagulls, cormorants, puffins, razor-bills, guillems [guillemots], and scrays [terns] of two sorts'.[15] The island, which lies off the north-east corner of Anglesey, is tiny, just 0.108 square miles, and its highest point is 60 metres (190 feet) above the sea. The term 'scray' comes from the terns' harsh calls; there are none breeding there now, but the island is still famous for its large colony (currently 700 pairs) of cormorants.

While Willughby's commonplace book had been ideal for recording his reading and philosophical notes while at Trinity, it was less suitable for recording his natural history observa-tions while travelling. It seems likely, therefore, that he also had other, smaller notebooks for specific topics or particular journeys. Although it is now lost, we know that he had a separate notebook for his visit to Priestholm since he mentions it in his other writ-ings. The notebook referred to could have been Ray's, but there's a tiny pointer that virtually confirms it was Willughby's, since he refers to it as 'PrestHolme Journy'. The clue is the capital H in the middle of 'Prestholme': an uppercase H in the middle of a word

Francis Willughby's writing of 'Prestholme' with an upper-case 'H'.

was one of the occasional quirks of Willughby's writing. Sadly, the location of the notebook itself – if it still exists – is unknown.[16]

Some seven miles inland from Anglesey, in the lakes 'hereabout, viz., at Llanberris [sic], Bettus, Festingiog, there is a fish called torgoch'. It was a name Willughby thought might mean 'salmon-like', but to Ray it was a reference to the fish's red belly. The *torgoch* was the charr and very similar to one Willughby had seen in Windermere a couple of years previously. Locally, the fish, described as 'blackish upon the back, red under the belly', was subject to some 'fabulous stories', including one in which 'three sons of the church brought them from Rome, and put them into three lakes'.[17] In fact the fish Willughby and Ray saw in Wales were three of what were once four distinct lake populations of charr. In his inimitable style Willughby described the fish, perceptively distinguishing it from similar salmon-like species, and, as a result, many years later it was named Willughby's charr in his honour.[18]

Llanberis was also where, on 26 May, they found a species of plant new to science, 'found on ye back of Snowdon not farre from Llanberis, in ye way from Carnarvan thither near an old tower'. This upland species, which now bears the name Small White Orchid (*Pseudorchis albida*), occurs widely across Europe, but remains rare (and declining) both in North Wales and elsewhere in Britain. When Ray published his *Historia Plantarum* in 1686, he recounted how he regretted not having made a full description of the species when he and Willughby first found it.[19]

On 29 May 1662 Willughby, Ray and Skippon took the somewhat perilous two-mile boat journey from Aberdaron off the tip of the Llŷn peninsula through swirling tidal rips to the beautiful island of Bardsey. Nipped like the waist of a wasp, the island boasts two crescent-shaped beaches, a mountain at one end and a green tail pointing southwards into the Irish Sea at the other. Once there Willughby and Ray saw at first hand why Bardsey was renowned as a holy burial place for saints, for near the ruins of an old church was 'a heap of dead men's skulls, and other bones of such votaries, as, for the sanctity of the place, had been buried there'.[20] Less macabre, they also saw 'puffins [Manx shearwaters] and sea-pies' (oystercatchers) and the beautiful blue spring squill flower 'growing in great plenty'. Bardsey was also the first island I ever visited, as an eleven-year-old boy; I saw no skulls, saints or their remains, but the seabirds and flowers would shape the rest of my life.

Returning to the mainland they travelled south, reaching St David's in Pembrokeshire on 5 June where, in the magnificent cathedral, they found 'diverse ancient monuments', reporting that 'the Welch have a proverb, that it is as good to go to St. David's twice, as to Rome once'.[21] Leaving St David's that afternoon, they rode eastwards along the coast towards Haverfordwest, passing Ramsey Island, named for the ransoms – wild garlic – that grows there, and 'saw at a distance, Scalme Isle [Skomer Island], but we went not thither'. It is a pity that they did not visit Skomer for it is (and has been for a long time) the most significant of the Welsh seabird islands, and where, following my Bardsey initiation, I have visited – to study seabirds – each year since 1972. As far back as the 1380s there is mention of 6s 8d received for the 'farm [harvest] of birds', which must refer to seabirds, such as the Manx shearwater that today has a breeding population there of over 200,000 pairs, and to the auks, guillemots, razorbills and puffins, which may have been similarly numerous in the 1600s.[22]

From Haverfordwest they rode south to Pembroke, visiting the exquisite tiny chapel at 'St. Gobin's [St Govan's] ... 'sacred to that saint' – built over the cave in which he lived and tucked away half-way down the cliff. Below it is 'a well, famous for the cure of all diseases'. St Gofan (his Welsh name) died in the sixth century and the chapel was built in the fourteenth. Intriguingly perhaps, neither Ray nor Willughby mention a series of spectacular sea stacks lying less than half a mile west of the chapel, famous today at least, and known as Elegug Stacks for the huge numbers of elegugs (guillemots) that breed there.

ᐟᐣᐟ

Seabird colonies provide some of the greatest of all wildlife spectacles. Smelly and noisy, the birds are abundant, conspicuous and can often be observed at close range. Since the sites where seabirds breed are traditional – the birds returning year after year – they are usually well known historically and not least because the adults, chicks and eggs were a source of food for local people. It is hardly surprising then that Willughby and Ray made a point of visiting several seabird colonies in their travels, and later included a specific section in the *Ornithology* entitled 'Of some remarkable Isles, Cliffs, and Rocks about England, where Sea-fowl do yearly build and breed in great numbers'.[23]

Their list comprises thirteen sites, although 'England' is a bit of a misnomer for they include the Bass Rock in Scotland, and Priestholm (Puffin Island), Bardsey Island, the Tenby cliffs and the islands of Caldey and St Margaret's, all in Wales.

In the previous year, during the late summer of 1661, John Ray and Philip Skippon – Willughby was not with them on that trip – had visited Bass Rock in the Firth of Forth. Ray's account in the *Ornithology* includes a long quote from William Harvey, physician to Charles I and discoverer of the circulation of the blood, who visited Bass Rock in 1633 during a trip to Scotland with the king for his coronation in Edinburgh. Harvey visited Bass Rock because he was fascinated by reproduction and felt that birds' eggs held the secrets to fertilisation. They did, but Harvey wasn't able to figure it out.[24] However, he found the mass of gannets on the Bass remarkable, as Ray and Skippon must have done. Evocatively, Harvey wrote 'such a number of birds there is flying over ones head, that like clouds they cover the skie, and take away the sight of the sun: making such a noise and din with their cries that people talking together nearhand can scarce hear one another'.[25] Somewhat less lyrically, Ray wrote 'We saw on the rocks innumerable of the soland goose [gannets]'[26] and that 'the young ones are esteemed a choice dish in Scotland, and sold very dear (1s. 8d. plucked). We eat [ate] of them at Dunbar ... the young one smells and tastes strong of ... fish' – unimpressed, I sense. Ray listed the seabirds that bred on the Bass, and despite the fact that at the date of their visit – 19 August – most of the auks would have finished breeding and left, they saw 'the scout's [guillemots] eggs, which are very large and speckled'. I suspect that the guillemot eggs shown to Ray and Skippon would have been taken earlier in the season by their guide: Ray's description is an understatement for not only are guillemot eggs large and speckled, they are remarkable for their bright hues and extraordinary pointed shape, features that Willughby and Ray later mentioned in the *Ornithology*.

The second seabird site in their list is the Farne Islands off the Northumberland coast. Ray and Skippon saw 'Farn Island [sic] at a great distance' during their 1661 journey, but didn't visit it. However, Ray did so during a later 'simpling trip' in July 1671 with the plant collector Thomas Willisel.[27] Ray lists the birds breeding on the Farne Islands: 'guillimets [guillemots], scouts or razorbills, coulternebs [puffins], scarfs [shags], Cuthbert duck [eider], annet [kittiwake], mire crow [black-headed gull], pick-mire [tern], sea-piots [oystercatchers], kir-bird [=? possibly a tern; kir-mew = common tern, kir being onomatopoeic, mew = gull], a sort of Columbus, less than a magpie, black and white, stands straight upright', and 'Gorges a fowl bigger and redder than a partridge'; this sounds like red grouse, and indeed a 'Gor' or 'Gor-maw' was the name for a red grouse, but the Farne Islands appear to be an improbable place to encounter this species.[28]

Once again, it seems clear that this list is at least partly second-hand and wholly confused. The 'sort of Columbus' is double counted here as the puffinet, which in the *Ornithology* Ray says 'argues to be puffins, but the description here given us of them (for we saw not the bird) agrees rather to be the Bass-turtle'. As if this wasn't difficult enough, the same species in the *Ornithology* is referred to as 'the sea-turtle' or 'Greenland dove'. In the text Ray writes: 'I guess this bird to be the same as with the puffinet of the Farn [sic] Islands ... and ... I persuade myself also, that it is the same with the turtle-dove of the Bass Island.' He adds: 'Why they call it a Dove or Turtle I cannot certainly tell. It is indeed about the bigness of a turtle [dove], and lays (they say) two eggs at once like them, and possibly there may be some agreement in their voice or note.'[29] From the description this is the black guillemot – not now known to breed on the Farne Islands – but one of only a few seabirds to lay two eggs (as doves do).

One seabird location mentioned in the *Ornithology* is the somewhat cryptically named 'Pile of Foudres'. 'Pile' refers to a castle, an informal term still in use today; and 'Foudres' is either an old spelling or misspelling of Fouldrey, meaning fowl (or bird) island, about four miles from Dalton on the Lancashire coast. This is known today as Walney Island, a low-lying, windswept island on which large numbers of herring and lesser black-backed gulls now breed.Although Willughby and Ray were in the vicinity of Walney during their 1660 journey, they may not have seen it for themselves. Instead they seem to have passed some fifteen miles to the north of Ravenglass, from where they took a boat to the Calf of Man.

Seabird colonies listed by Willughby and Ray in the *Ornithology*.

1. Bardsey Island	visited by Willughby and Ray 1662
2. Bass Rock, Scotland	visited by Ray 1661
3. Caldey Island, Wales	visited by Willughby and Ray 1662
4. Calf of Man	visited by Willughby and Ray 1662
5. Farne Islands	seen by Ray from the mainland in 1661 and visited by him 1671
6. Fouldrey, Lancs. (aka Walney)	Willughby and Ray came close in 1662 but did not visit
7. Godreve Island	visited by Willughby and Ray 1662
8. Herm, Guernsey	not visited by Willughby or Ray
9. Lundy Island	not visited by Willughby or Ray
10. Priestholm, Wales	visited by Willughby and Ray 1662
11. Scarborough	visited by Ray 1661
12. Scilly Isles	not visited by Willughby or Ray
13. Tenby, Wales	visited by Willughby and Ray 1662

Note: I have used modern spellings of the place names. Also mentioned in the *Ornithology* are Ramsey Island and Skomer Island, Wales, but with no reference to seabirds; many others, including the colonies on the Flamborough headland, Yorkshire, are omitted.

Their list of British seabird colonies must have been included in the *Ornithology*, I presume, for 'completeness' despite its incompleteness (many colonies in Scotland are not mentioned), even though they had visited so few of the sites themselves.

Willughby and Ray's travels in England and Wales, 1662.

Continuing eastwards they arrived at Tenby where, during the first week of June 1662, they were excited to discover a wonderful diversity of fish in the harbour. Ray lists no fewer than thirty-two species, including conger eel, sprat, thornback ray and belone (garfish). The fact that these are so meticulously listed suggests that Tenby was something of a piscatorial treasure trove; a wonderful opportunity to compare and contrast the different species in the flesh – providing crucial information that Willughby would later use in his *History of Fishes*.[30]

Across a narrow channel from Tenby lies Caldey Island, with its three chapels, and 'in a little island [St Margaret's], between that and the main land, great plenty of fowl, the same as breed in Prestholm. In one part of the island the puits and gulls and sea-swallows' nests lie so thick that a man can scarce walk but he must needs set his foot upon them'. Ray points out that 'The sea-swallows they there call spurs, and the razor-bills are called *Elegugs* … This name elegug some attribute to the puffin, and some to the guillem [guillemot]; indeed they know not what they mean by this name.'[31]

After delighting in the seabirds of Caldey and St Margaret's Island, Willughby, Skippon and Ray mounted their horses and continued eastwards. At Laugharne, seven miles south of Carmarthen, they forded the river Cywy and in bright sunshine headed for the town of Kidwelly (Cydweli). Riding three abreast along the beach they reflected on the fact that the guillemots, razorbills and puffins were so obviously distinct, yet to the men of St Margaret's, who referred to them all as 'elegugs', they obviously seemed the same. I imagine it must have been after a conversation such as this that Willughby and Ray decided that natural history needed an overhaul. Different species known by the same name, the same species known by mul-tiple names – what a mess! The new science, with its emphasis on objective description, quantification and the avoidance of ambigu-ity, provided a way forward. Seeing things for themselves, whether a shearwater chick on the Calf of Man, the multitude of fishes at Tenby, or the moonwort at Sutton Coldfield, was what gave them the clarity of vision to create a new natural history. It was a decision that brought their objectives into sharp focus, and one that would shape both their lives and their legacies.[32]

Local names of some birds used or mentioned in the *Ornithology*, whose modern common and scientific names are given in Appendix 4.

Ars-foot	Gorcock
Bald buzzard	Green Plover
Bastard Plover	Greenland Dove
Black-cap	More-buzzard
Bohemian Chatterer	Ox-eye
Cock of the Mountain or Wood	Pool Snipe
Common Grosbeak	Puffin of the Isle of Man
Copped Douker	Puttock
Coulterneb	Pyrag
Daker-hen	Rock Ouzel
Didapper	Skout
Dun-diver	Small water hen
Fern-owl	Solitary sparrow
Flusher	Water Ouzel
Gid	Witwall
Glead	Woodspite

As Willughby, Skippon and Ray arrived at Aberavon, near present-day Port Talbot in South Wales, on Thursday 12 June 1662, two significant events occurred. The first involved an encounter with an unusual bird said to breed nearby, and 'supposed to be the *Hamantopus*', whose description fits a black-winged stilt. They must have seen both a live bird and a specimen, for their description is very detailed: 'the first five or six feathers of the wing above of a dark or fuscous colour, near black, underneath more light or don-nish ... the legs long and red ... these she stretches backwards in flying, which makes amends for the tail; it makes a piping noise.'[33] Its later generic name *Himantopus* comes from the Greek mean-ing 'strap-legged' or 'long-legged'. Curiously, when Willughby and Ray came to write their account of this species in the *Ornithology*, they either forgot that they had seen it or were unable to match it with what two well-known sixteenth-century naturalists, Conrad Gessner or Ulisse Aldrovandi, had written about it, saying: 'To say the truth, it hath not been our hap as yet to see this bird.'

The second event involved Willughby interviewing a local person as part of the language project proposed by John Wilkins. Realising that they would soon cross back over into England, this was Willughby's last opportunity to gather some Welsh words. Not any old Welsh words, but words as spoken by a Welsh speaker. Wilkins had suggested to Willughby that he compare the words used to identify or signify particular things in different languages and to this end Francis prepared a list – a prompt list – of English words from which he then obtained their equivalent from a Welsh speaker. Not speaking Welsh himself, of course, Willughby wrote the Welsh equivalents in a phonetic manner. Whether Wilkins had suggested this approach, rather than letting the Welsh interviewee write his own 'words', we don't know, but by writing them himself, Willughby created an objective and consistent way of collecting the relevant information.

Far from being a random selection of words, the prompt list was derived from an earlier work on linguistics by the Scot George Dalgarno, a colleague of John Wilkins, and from Wilkins himself. The list started, inevitably perhaps, biblically (God, angel, heaven), moving on to celestial things (sun, moon, stars), the body (hair, skin, nail, eye), and then to animals and plants (bird, beak, wing, feather, fish, leaf, seed). By using the same prompt list of these carefully selected English or Latin words wherever they were, and by writing down the speakers' responses in a phonetic, and hence neutral, manner, Willughby and Wilkins were able to make objective comparisons between different geographical regions, and subsequently different countries. It was exactly like studying birds: examining their plumage and anatomy to identify their characteristic marks in order to classify them.

Prior to this, those interested in words or dialects had simply made collections of anecdotes, assuming that regional variation in the use of words was the result of linguistic ignorance or laziness, and that any variation occurred at random. The approach of Wilkins and Willughby was fundamentally different. First, because they assumed that variation existed systematically across regions, and second, because their method was rigorous and objective.

Although Willughby's work on words was never published, linguists subsequently recognised it as a major advance in the study of language.[34]

After crossing the border back into England, Willughby, Skippon and Ray headed towards Gloucester to see its magnificent cathedral, arriving on the evening of 14 June 1662. I imagine them riding up the city's main street as it is getting dark, finding an inn, stabling their horses, getting something to eat and tumbling into bed. After breakfast the following morning they walked across to the cathedral's statue-laden entrance. One of the few such buildings to be spared by Henry VIII (possibly for personal reasons), the cathedral is an architectural masterpiece. Inside, they admired the twelve tiny chapels, the nave's bulky columns, and the ornately carved choir with its curious misericords. Walking through the presbytery they stand together to gaze in amazement at the vast vaulted ceiling and the enormous multi-coloured stained-glass window. Their clergyman guide directs them towards a small spiral stone staircase from which, thirty feet up, they emerge onto the north ambulatory. Close by, but almost hidden, is the entrance to a small and awkwardly angular passage connecting the north ambulatory with the south, running behind the great window. Known as the 'whispering gallery', this is the object of their visit. Intrigued, Willughby asks their host to demonstrate. And so, positioned at either end and separated by some eighteen metres of tomb-like tunnel, his host whispers while Willughby waits. And sure enough, in the cathedral's silence, the whispered words ricochet along the passage walls, maintaining their clarity and accruing an extraordinary amplification. In his commonplace book Willughby wrote: 'Ye voice is hard [heard] as plane [plain] or planer [plainer] in any part of ye vault as at ye ends.'

What was the secret of this curiously kinked corridor? Intrigued and confident that the new science could provide a logical explanation, Willughby set to work enthusiastically measuring the gallery's dimensions, scribbling numbers and ideas in notes that are hard to interpret:

Y^e first [side] E[ast] side is 7/8 allmost, /
ye second 1 ½, ye third 5, ye fourth /
to ye first Windore 1 1/4 , thence to y^e /
Doore 3 ¼, ye doore 7/8, to ye furthest /
windore 2 ½, to ye Angle ½, /
ye fifth 5 ½, ye sixth 3. /
(it is Broad at Both Ends a little /
less then a yard.) Y^e Breadth /
quite thorough is something lesse /
then a yard. /
Y^e Height measuring from ye middle /
Angles is 2 & an Inch ye W[est] side, but /
ye E[ast] wants a little, ye roofe shelving /
downwards. /
there are 8 windores ye n[orth] side of y^e /
Doore & 6 by S[outh] side. /
[uncertain cancellation] at ye Ends it is a little /
Higher & ye roofe not so shelving.[35]

When I visited the gallery, I examined the scratches on the passage-way walls, wondering whether any were Willughby's, made as he marked out the dimensions. Probably not, but I liked the idea, and I searched in vain for his initials amid the dense graffiti of names and dates.

Had the gallery been built with whispering in mind or was its effect an acoustic accident? Willughby must surely have asked himself this, as I did after my visit. The diarist John Evelyn thought it was deliberate, and after experiencing the gallery in July 1654, wrote: 'The whispering gallery is rare ... and was, I suppose, either to show the skill of the architect, or some invention of a cunning priest, who standing unseen in a recess in the middle of the chapel might hear whatever was spoke at either end.'[36]

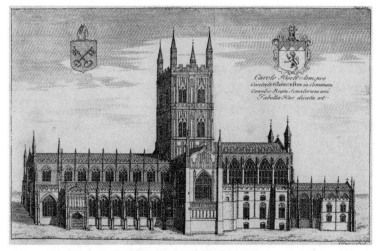

Gloucester Cathedral (1710).

I suspect it was an architectural fluke, with the narrow passage merely a convenient byway between one side of the east wing and the other. A plan drawn in 1807 confirms this, showing the gallery following the cathedral's contours.

The gallery being a curiosity long before Willughby's day, I wondered whether he was asked to pay to experience it, as I was. He ends his scribbled account by asking 'whither any entrie or vault would not doe [do] as wel'.

There's nothing more on this in the commonplace book, but those notes exemplify Willughby's inquisitiveness and interest in all that was curious and, crucially, his belief that all puzzles were amenable to interpretation. How many other scribbled notes, in other notebooks now lost, were there?

᛭

Shortly after leaving Gloucester, Willughby separated from Skippon and Ray to travel alone. He later wrote to Ray:

I met with several adventures in the remaining part of my journey after I left you; and among the rest with one very lucky one, of a new discovery of medals. You may remember the day we parted I had intended to have gone to Cirencester, but hearing ... of a great deal of treasure that was found in a field, I ... diverted my course thither. The field was near Dursly [Dursley] ... where I found above forty people digging and scraping; and bought a great many silver medals of them, and one incomparable fair one of gold, that had been found a little before. The whole history of how these came to be discovered, I shall reserve till I see you.[37]

Willughby continues, in his letter: 'I thought to make strict enquiry after the snap-apple bird, but falling very sick at Malverne [Malvern] was forced to give over all.' So he got his silver and gold coins, but illness prevented him from seeing the mystery bird, which it turns out was the crossbill. This species occasionally appears in large numbers when its normal food (pine seeds) fails; it then sometimes damages apples to gain access to and consume their seeds – hence its local name.[38]

The collecting of old medals and coins for their cabinets was fashionable among virtuosi. It was an affordable way of connecting with antiquity and inasmuch as it heralded – eventually – a 'new archaeology', it was part of the new science. And, just like natural history artefacts, coins and medals could be categorised and classified.[39]

After returning home and recovering, Willughby took himself off to watch birds in Lincolnshire. In a letter that Ray sent to Peter Courthope in February 1663 about the identity of the shoveller duck, he writes: 'Mr Willughby in one of his letters to me, containing a catalogue [list] of what foules [birds] he gate [watched] in Lincolnshire the last summer, hath one which possibly may be the same with this [i.e. the same as the shoveller Courthope had asked Ray about].' We know nothing more about this particular journey made by Willughby in the summer of 1662, but it confirms his growing interest in ornithology for Lincolnshire at that time was a vast wetland with a huge abundance and diversity of birds. It was also an area inhabited by those who made their living trapping

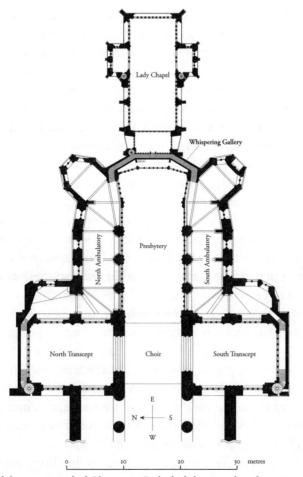

Lady Chapel

Whispering Gallery

North Ambulatory

Presbytery

South Ambulatory

North Transcept

Choir

South Transcept

E

N ← → S

W

0   10   20   30 metres

Plan of the eastern end of Gloucester Cathedral showing the whispering gallery.

waders, plovers, ducks and rails for human consumption, providing excellent opportunities for someone like Willughby to examine birds in the hand.[40]

With Willughby gone, Ray and Skippon continued from Gloucester into the West Country. Near Padstow they watched 'great flocks of

Cornish choughs' and learned about the cruel way local fisherman caught gannets 'by tying a pilchard to a board, and fastening it so that the bird may see it, who comes down with so great swiftness for his prey that he breaks his neck against the board'. From St Ives they visited Godreve Island where they again saw a large number of seabirds, remarking that 'Here they call the puffins popes and the guillems [guillemots] kiddaws.' At Penzance they saw and described more fish – and as with the birds, they commented on the different names used for the same species: the 'tub-fish which is no other than the red gurnard', wrote Ray, and 'tomlins, which are nothing but a young codfish'.[41]

On reaching Land's End, Ray and Skippon turned and headed for home, arriving in London in mid-July, where their journey terminated. If he didn't already know, this must have been when Ray discovered that the new Royalist Parliament required all ministers to sign an oath of loyalty: the Act of Uniformity.

The Act meant that anyone holding office in the Church of England – like Ray – now had to follow the form of prayers and ceremonies as laid down in the Book of Common Prayer and whose use was now compulsory in church services. This was 'one of the measures enacted by the Cavalier Parliament to secure the authority of the restoration regime of Charles II'.[42] The Act undermined Ray's religious beliefs and together with some 2,000 others, he refused to sign and was thrown out of the Church, losing his Fellowship at Trinity College and with it his livelihood. It also meant forfeiting any further involvement as a clergyman. As a result, on St Bartholomew's Day, 24 August 1662, at the age of thirty-five, Ray found himself unemployed.

Although this seems like an abrupt and worrying change in his circumstances, Ray had been thinking for some time about leaving Cambridge and taking a position in a private school. A possible reason for this is that after the Restoration in 1660 and the return of the previously evicted Royalist Fellows to Trinity, who – crucially – assumed their original positions in the college pecking order, Ray must have recognised that his chances of working his way up that hierarchy were now much reduced.[43]

Historiated initial 'C' showing Charles II – from the charter accompanying Gloucester Cathedral's sealed copy of the Book of Common Prayer 1662.

Nonetheless, concerned about what the Act might mean, he wrote to Peter Courthope:

August 24 has passed by now and I have not returned to Cambridge: consequently, the die is cast; behold I have been ejected from the fellowship [at Trinity] without any rights to return; for me, therein nothing more is [to be] sown or reaped; and I must seek a new way of life in some other direction.[44]

Ray may also have been concerned about the atmosphere at Trinity. Prior to 1660 he had been part of a circle of like-minded colleagues who included John Wilkins, Isaac Barrow, Henry More and John Cudworth, all of whom had been actively engaged in the new science that included animal dissection, experiments in 'chymistry' and botany. Following the Restoration, such activities seemed – to Ray at least – less popular and less exciting.

With little option but to 'cast himself upon Providence and good friends', Ray returned to his mother's house at Black Notley in Essex.[45] Willughby at Middleton, meanwhile, had been thinking hard about their earlier conversation regarding the makeover of natural history. Having decided that such a project, including a protracted research trip to the continent, was feasible, Willughby wrote to Ray:

> I am likely to spend much of my life afterwards in wandring or else in private studiing at Oxford, having but little heart to thinke of settling or ingaging in a family. I shall bee verie glad of your constant company and assistance in my studies and must again desire you by no means to part with your bookes.[46]

This was a lifeline, both intellectually and financially, I suspect, for Ray, and the beginning of a new career for both himself and Francis Willughby.

Buzzing with excitement at the prospect of this bold venture, Francis Willughby attended his inaugural meeting of the Royal Society in London on the first day of October 1662. He had carried with him for the 'show-and-tell' that was the Society's format, a single snake's egg, which was duly opened for the benefit of an audience that included Robert Boyle, Christopher Wren, Robert Hooke, John Evelyn, Christopher Merret, George Ent and John Wilkins. The egg, which I suspect was that of a grass snake, contained a well-formed embryo within a few days of hatching, whose yolk sac the Fellows likened to a placenta. Also of interest was the presence of 'two little protuberant parts near its tail, taken by some for a *geminus penis*', that is, the snake's two hemipenes. It was already known that snakes and lizards possessed a pair of penises, but it was not known then that these develop outside the snake's body during embryogenesis and are drawn inside the cloaca only on hatching. Stimulated by the little snake, the audience went on to discuss the different modes of development with vipers, as was well known, which brood 'their eggs within their bellies, and bring forth live vipers', whereas others – such as the grass snake – 'lay their eggs in dunghills, by the heat of which they are brooded'. One (unidentified) 'member of the society added, that he had seen a snake lie upon its eggs in the manner, that a hen sits upon hers'.[47]

Willughby must have been delighted by the response to his initial performance. The fact that he attended three consecutive weekly meetings of the Society that month indicates that he was in London for the duration, presumably seeking advice from other Fellows and securing the necessary papers for the forthcoming continental journey.[48]

His time at the Royal Society that month allowed Willughby to familiarise himself with, and absorb, its aims, objectives and boundaries, beyond merely seeing things for himself. Still finding its feet, the Society made a public statement that its business was to 'improve the knowledge of natural things and all useful arts, manufactures, mechanic prectises, engynes and inventions by experiments … and … explicating all phenomena produced by nature'. At the same time the Society made it clear that there would be no 'meddling with divinity, metaphysics, morals, politics, grammar, rhetoric or logick'. It also urged its members when reporting their findings to use plain language, 'bringing all things as near the mathematical plainness, as they can'. The Society was also obsessed with keeping meticulous records, partly as a way of advertising its seriousness, but also as a way of ensuring its own permanence.[49]

Willughby and his travelling companions were to keep very careful records during their continental journey. It is ironic that Willughby's are now lost, but we are fortunate that both John Ray and Philip Skippon kept extensive notes, and it is their accounts that allow us to relive those exhilarating days.

Continental Journey: The Low Countries

After months of planning, on Friday 17 April 1663, John Ray and Philip Skippon mounted their horses at Leeds Castle in Kent and rode the thirty-five miles to Dover to meet up with Francis Willughby, Nathaniel Bacon and two servants. At two o'clock the following afternoon they boarded the packet boat, having paid 'five shillings a man for [their] passage, and five shillings for the use of master's cabbin [sic]'. By eight in the evening they were becalmed and 'forced to lie two leagues short Calais till the morning', arriving eventually at dawn. There they were met by two French boats and forced to pay 'three-pence a-head to the master of the ferry ... but before they would let us ashore, after much wrangling with those brawling sharking fellows, we were forced to give them six-pence apiece'.[1]

So began the first day of their monumental journey through Europe. Why monumental? Not just because of the distances they were to travel, on foot, on horse and mule, by coach, boat and barge, but because of the sheer magnitude of their undertaking. Their aim was to see, document and absorb everything – including birds – they encountered. The success of this carefully planned educational adventure depended upon diligent, sustained reporting.

As an undergraduate Francis Willughby had enthusiastically imagined himself enjoying some form of educational travel. His father and many of his Trinity colleagues had journeyed to the continent, and Willughby must have seen and heard first-hand how stimulating, valuable and occasionally frustrating it could be. As early as the 1500s travel was widely acknowledged as enhancing, complementing and completing a young gentleman's education. Indeed, it was almost expected, and as Willughby's commonplace book makes clear, he had read widely about foreign travel while at Cambridge.

James Howell's *Instructions for Forreien Travell* published in 1642 must have been among Willughby's most important sources of inspiration and information as he and Ray made their preparations during the winter of 1662–3. Educated at Jesus College, Oxford, Howell was a well-travelled historian, linguist and prolific writer, and 'one of the earliest instances of a literary man successfully maintaining himself with the fruits of his pen'.[2] A major theme in Howell's book was the way foreign travel enriched the mind, a notion that Willughby undoubtedly relished. Howell also provided vital information on the value of letters of introduction, the costs of travel, which countries were barren or fruitful, crossing the Alps, and significantly the difficulties of travelling in Spain, a notion that seems to have captivated Willughby.[3]

Above all else Howell emphasised 'the preeminence of the eye' – the true value of seeing things for oneself – and this must also have struck a chord with Willughby and his Royal Society colleagues. You might be forgiven for thinking that Howell had written the seventeenth-century version of a Lonely Planet Guide: in a way he had, but in contrast to today's travel guides one has to work hard to extract the relevant information, for brevity and clarity were not Howell's forte. But Willughby would have been used to that style of writing – despite it being so different from his own. The true value of travel was – and still is – beautifully summed up by the Flemish travel writer Justus Lipsius, who referred to those content to stay at home as being like 'sillie birds cooped up in a pen'.[4]

Reading about travel was, literally, only the beginning. The most important challenge was the itinerary, made easier by the fact that in the mid-1600s gentlemen travellers had already established two well-trodden tracks across the continent – referred to as 'the circle'.

The first, through France via Paris and Montpellier, ended in Italy; the second reached Italy via the Netherlands, the major cities of Germany, the Czech Republic and Austria. The Iberian Peninsula, North Africa, Russia and Turkey were *not* recommended.[5] Where and when Willughby's party went was dictated partly by the need to avoid any trouble arising from a slightly frosty phase in Anglo-French relations at that time,[6] and partly by who and what they wanted to see. Unlike many of their predecessors, Willughby and his companions also needed to be at particular locations in the appropriate seasons to see particular plants and animals. For these reasons they rejected 'the circle' and devised their own route instead.

Contacts were crucial, and scattered across Europe were numerous Trinity College men who formed a loosely connected intellectual brotherhood into which Willughby knew he and his companions would be warmly received. There were also particular individuals who they were keen to meet, such as Christiaan Huygens, one of Europe's most eminent scientists, famous for his studies of astronomy, optics and games of chance.

What to see? High on their list were the great universities of Europe: Leiden for its expertise in medicine, philosophy and natural history; Padua for anatomy; and Montpellier for its botanical and pharmaceutical experts. Other priorities were museums and collections, most of which were in private hands, and many of which, like the impresario Athanasius Kircher's 'closet of raritys' in Rome, were well known to continental travellers.[7]

Letters of introduction from eminent and trustworthy individuals were essential. Willughby, for example, carried a letter of recommendation from John Wilkins to Huygens; and knowing – contrary to advice – that he would travel through Spain, he had also obtained letters of recommendation written in Spanish from his colleague Charles Baworth.[8]

Passports were obtained from Whitehall and cash was acquired from London merchant houses. Recognising that dissection was to be an essential part of their endeavours, Willughby and his colleagues carried guns and sets of surgical instruments – scissors, lancets, knives, forceps, hooks, probes and seekers – all possibly contained in a small chest. They also needed boxes to hold insects and bird specimens, and presses to preserve their botanical specimens.

Today's travellers interested in natural history might carry a compact field guide either as a book or on their iPad, but of course nothing so convenient was available for Willughby's party. There were books – those of their predecessors – but these were inconveniently bulky and they cannot possibly have carried them all. Nonetheless, they probably did take some and there's a hint that one of these was Guillaume Rondelet's book on marine animals, *De Piscibus*.[9] It is hard to imagine them being able to achieve what they did en route without access to the earlier literature, and I suspect they consulted the standard texts whenever they had access to the libraries of people they visited. In addition, they probably carried some of the travel guidebooks mentioned above.

Notebooks were indispensable. Unfortunately, all of Willughby's have been lost, but we know that John Ray used small, slim, handmade notebooks comprising sheets measuring 10 x 15 centimetres stitched into oatmeal wrappers.[10] Between Willughby, Ray, Skippon and Bacon there must have been many such notebooks. Essential too would have been a 'system', most obviously a different notebook for each topic: one for their general observations, one for birds, one for fish, for language, antiquities, trades and so on. Willughby's and Ray's observations were so diverse that without careful organisation and a conscious effort to keep records separate their notes would soon have degenerated into a mess. As well as notebooks, they also carried with them prepared questionnaires, including word lists such as those they had used previously in Wales.

Trained as many of us are today in note-taking at school and university, as well as using digital folders on our laptops, it is easy to forget that Willughby and his colleagues would be treading unfamiliar paths in terms of organising their accumulated observations. Luckily, one or two of their predecessors had anticipated this potential problem and had published suggestions for classifying information. The Swiss physician Theodor Zwinger paved the way in 1577 with a book entitled *Methodus Apodemica* – literally, 'methods of travel', but specifically scholastic travel, exactly what Willughby and Ray were doing.[11]

Zwinger's approach was to construct tables that allowed travellers to place information in relevant categories, greatly facilitating the subsequent analyses of that material – an eminently sensible suggestion. Zwinger's student Hugo de Bloot (aka Blotius), a librarian, scholar and natural historian, later developed a travellers' checklist of what to record: geographic location, money, buildings, churches, rituals and so on. And even if Willughby and Ray didn't take Bloot's checklist with them, it may well have inspired them to construct their own.

Finally, there was the issue of how one should conduct oneself while travelling. The answer, of course, was with decorum and critical acumen, as well as discretion when discussing religious and political matters. It is highly unlikely that all those undertaking the Grand Tour of the continent behaved in this way, but because Willughby and company's tour was a fact-finding mission, appropriate etiquette undoubtedly helped to ensure their success.

Crossing the English Channel entailed more than simply a change in language and religion. It also involved a massive ten-day change in calendar date, for France had been among the first countries to switch, in 1582, from the Julian (Solar) calendar to the Gregorian calendar. Proposed by Pope Gregory XIII, the change was motivated by the need to correct for the drift in dates caused by the fact that it took 365 days, 5 hours and 48 minutes, rather than exactly 365 days, for the earth to circle the sun. The switch in calendar, which facilitated the realignment of Easter with the vernal equinox (as it was meant to be), corrected for the accumulated drift through the 'loss' of ten days. It also resulted in potential confusion, some of which is alleviated by the terms 'Old Style' and 'New Style' when referring to the Julian and Gregorian calendars respectively. Further potential for misunderstanding was caused by different countries switching calendars at different dates, with Great Britain adopting the Gregorian calendar only in 1752. Willughby and his colleagues, however, stuck assiduously to their familiar Julian calendar throughout their travels.

The day after arriving in Calais they set off for Dunkirk, beginning as they intended to carry on by soaking up all that was different. Perhaps surprisingly, their journey started with a visit to some English nuns: 'They spake very civilly to us, and told us they were in number 44. They live very strictly, and never see the face of any man; the bars were of iron that we discours'd through.' At another convent of English nuns the same day, Willughby and his three friends had 'the freedom to see and discourse with the ladies; about five or six giving us the entertainment of their company through an iron grate'. Down at the quay they examined some fish, noting the 'marner' (whose identity remains a mystery), another 'some call'd tench', and a third, the 'potshoest, i.e. *Scorpoena bellonij* [= Scorpion fish *Scorpaena* spp]'.[12]

In the palace gardens at Brussels on 30 April were two captive eagles, an ostrich and two white Muscovy ducks. The ducks may have been noteworthy because the wild Muscovy duck, native to Mexico, Central and South America, has glossy dark-green plumage and was first brought to Europe in the sixteenth century. The domesticated, white-plumaged birds were sufficiently unusual to secure themselves a place in the *Ornithology*. In a later publication John Ray explained the name by saying that it is nothing to do with 'Moscow', but pertained to the bird's musky odour – the function of which remains unknown.[13]

They visited the university at Louvain, recording the way the students were taught and that they wore gowns and square caps, noting also that when the students enrolled at this university they had to 'swear their belief of all the doctrines of the *Romish* church'. The city itself, they felt, 'for trading and wealth is much decayed since the Low-Country [Eighty Years] wars'. Two days after leaving Louvain they were in Antwerp where Ray listed the many rare plants in the garden of the priest Franciscus van Steerbeck.[14]

There were many stalls 'well stored with fish of several sorts' at Machlin (Machelen): 'we saw the Vinder-fish or Vintz [again, unknown], Cods, piscis Mai, i.e. Alosa sive clupea [allis shad], Barbles, Holybutt [halibut: holy because it was eaten on religious holidays, butt = flatfish], Hootes, i.e. Oxyrhyncus [sturgeon?], and Eless [= eels?]'. In a druggist's shop, they examined a preserved armadillo, a dried sturgeon and a 'little square fish having a round mouth, two horns before on the head, and as many on the tail' (possibly a long-horned cowfish *Lactoria cornuta*, from the Indo-Pacific – and undoubtedly a dried specimen). In the same shop were preserved crocodiles and alligators and two 'horns', one of which was over eight feet long. These were the tusks (an upper canine tooth) of the narwhal, widely believed at that time to be the horn of the unicorn, and because of its alleged magical powers a must-have item for any cabinet of curiosities. Although Olaus Magnus, archbishop of Uppsala, published an image of a fish-like creature with a single long horn in the mid-1500s, it was only in 1636 that the Danish physician, Ole Worm, famous for his own Wunderkammern (wonder room or cabinet of curiosities), discovered the true identity of the unicorn's horn, which he did by obtaining one still attached to a skull of what was obviously a whale. Ray knew this too, writing that it was 'the horn of a fish of the cetacean kind ... not the horn of a quadruped, as is vulgarly but erroneously thought'.[15]

In a different druggist's shop they witnessed another Arctic curiosity: a 'Greenland man in a boat'.[16] This was a copy, as Skippon knew, of 'that which hangs up at Hull in England'. This particular curiosity had its origins off the coast of Greenland in 1613, when the crew of an English vessel, the *Heartsease*, rescued an Inuit found exhausted at sea in his kayak. They brought him aboard and tried to revive him, but he died three days later. On their return, an effigy of the man seated in his kayak, and in his original clothes, was displayed by Trinity House in Hull.[17]

On 15 May 1663 the party took a boat to Middelburg and were searched, 'as were all vessels going to and from Antwerp', for security reasons.[18] There in the town hall they saw two black eagles 'shut

Drawing of the Inuit man in his kayak rescued by the Heartsease *in 1613.*

up in a cage', which the town's charter apparently obliged it to keep.[19] Ray later described these as 'double the bigness of a raven, but lesser than the pyarg [white-tailed sea-eagle]'; and said 'of the place of this bird, its food and manner living, building its nest, eggs and conditions, etc., we have nothing certain'.[20] Given that they did not know what species it was, it is hardly surprising they knew nothing about its biology. Their 'identification', it seems, was based on Aldrovandi, who mentions 'black eagles' in his books; but, in all probability, the Middelburg eagles were golden eagles.

Also in that town they visited the house of someone named 'Cliver' (presumably Kluijver), who showed them some of his rarities, including 'sea-horses teeth' (in all likelihood the tusks of a walrus, once known as the seahorse), a whale's penis (dried, I presume), more 'unicorn' horns, a sea wolf (possibly a dried specimen of the northern wolf fish *Anarhichas denticulatus*), a sea porcupine (possibly the tropical porcupine fish *Diodon holocanthus*, or the spiny puffer fish, both of which when dried make spectacular museum pieces), an ostrich egg with faces carved on it, twelve dodecahedrons of ivory one within another, a circumcision knife made of a blueish stone, and 'dragon's teeth, i.e. the petrified teeth of a shark'.[21]

From Middelburg they travelled on to Bergen op Zoom from where on 21 May they hired a 'waggon [sic] drawn by three horses abreast, which carried us ... to a village called Sundert'. It was here that they shot a red-backed shrike. Much later when writing the *Ornithology*, Willughby and Ray refer to this as *Lanius tertius*, 'called in Yorkshire the flusher'.[22] Shrikes would have been easy to shoot because of their habit of sitting out in the open on the top of a bush, and they killed another 'between Heidelberg and Strasburgh, about a village called Linkenom'. This one was 'the greater butcher bird' or great grey shrike, of which they comment: 'We are told, that it is found in the mountainous parts of North England, as for instance in the Peak District of Derbyshire' – which is true, but only as a winter visitor.[23] Two other types of shrike are named and described in the *Ornithology*, but both are woodchat shrikes, the first an immature bird with 'the breast elegantly variegated with the like black semicircles, almost after the manner of the wryneck'; the other is clearly an adult bird 'whose head was a lovely red'. We also learn from their account that Ray later saw and described another woodchat shrike in Florence – almost certainly from the bird market – and 'Mr Willughby also described another killed near the Rhene [Rhine] in Germany.'[24]

From these descriptions we get a clear sense of Willughby's eye for detail. Of the great grey shrike, he says 'the nostrils are round, above which grow stiff black hairs or bristles'. We now refer to these as rictal bristles – modified feathers – whose bases contain numerous sensory nerve endings. Of the red-backed shrike, Willughby commented on the notch on the upper mandible of the bill; a feature found in few other birds.

'About six in the afternoon we took our seats in the passage-boat, somewhat like our pleasure-barges on the Thames ... and by one horse were drawn in two hours time to Delft.'[25] This gentle, undemanding form of travel must have provided a welcome change from their horses and an opportunity for the four men to chat, make notes and discuss what they'd seen so far, and what they hoped to see at their next stop.

The cabinet of curiosities owned by the Delft apothecary Jean van der Mere (Jan van der Meer) was packed with intriguing zoological specimens, including a civet cat, a moose with enormous palmate antlers, the skin of a rattlesnake and an elephant's tail that Ray considered 'a very small thing considering the bigness of the animal'.[26] Also in Delft they found a 'chirurgeon's' (surgeon's) anatomy theatre stuffed with exciting specimens: the mummified leg of a man, 'a flying cat or squirrel with membranous wings and tail', the head of an elephant, the skull of a babirusa, *Babyrousa babyrousa* (an Indonesian wild pig whose extraordinarily long tusks curl upwards from the lower jaw, through the upper lip and towards the eyes), and a starfish with five radii on a convex shell that Skippon's sketch allows us to see is the 'test' or skeleton of a sand dollar (an echinoderm). In Ray's notes on their visit (and comparing his with Skippon's it seems they shared but didn't completely duplicate information), he mentions a 'Soland-goose out of Groenland', which may be a northern gannet; several sorts of hummingbird; the dried head of a hornbill (said to be 'worth twelve florins in Amsterdam'); a feather garment from the Straits of Magellan; the egg of a cassowary; and the eggs of an Indian goose – whatever that refers to.[27] This final example makes the point that as exhilarating as it was for Willughby and his colleagues to see and handle such unusual objects, with little or no information or context – and indeed, a high risk of misinformation – it must have been frustratingly difficult to know what to make of them. In addition, unless they made notes as they went around these collections or were given a catalogue or list, it must have taxed their memory each evening to record everything they had seen during the day.

From Delft they travelled once again by horse-drawn boat to The Hague, arriving at Leiden three hours later. At the university there they were interested in what the students were reading and no doubt drew comparisons with Cambridge.[28] Leiden's physic garden comprised 'a square of less than an acre of ground, but well stored with plants, of which there have been at sundry times several catalogues printed'.[29] In the anatomical theatre they found 'preserved many skeletons of men and beasts, skins of beasts, parts of exotic animals, and other rarities'. They saw horned beetles from the East Indies, petrified mushrooms (=?), the skin of a '*Tartarian* [Tartar] prince who ravished his sister', an anteater 'bigger than an otter, having a very long snout, long crooked claws, coarse bristly hair and a long brush tail'. This was obviously the giant anteater *Myrmecophaga tridactyla* from Central and South America. Even today the size and shape of this extraordinary animal seems improbable, so Willughby and his companions must have been amazed by it.[30]

In Leiden they took the opportunity to attend some lectures at the university, including one in which the insect that 'makes the Indian cochinele' was described – the highly prized carmine dye cochineal, extracted from a Mexican scale insect. Beneath the anatomy theatre they saw the 'great skeleton of a fish we guess'd to be a whale', presciently noting that the shoulder blades were like those of a quadruped. Willughby and Ray knew that whales weren't fish, but there was a widespread perception that they were.[31]

A visit to Leiden's public library accompanied by a Mr Newcomen, minister of the English congregation, allowed them to examine the original papers of the famous historian Joseph Scaliger. It may have been Scaliger's interest in philology – the study of literary texts – that attracted Willughby and his party. It was then on to visit Dr Van Horne, professor of anatomy, who entertained them with 'great kindness and civility' and showed them many anatomical specimens, of which the most notable were the skulls of human embryos 'wherein were clearly discern'd the disjunction of bones which are afterwards not to be observed, the intermediate cartilages hardening to bone'. He also showed them the three tiny bones of the ear: the hammer, stirrup and anvil, and the 'bones found in the *glandula pinealis* of men, which were very small'. But Van Horne was mistaken; the pineal gland (located near the centre of the brain) contains no bones. The pineal is an endocrine gland shaped like a pine cone – hence its name – that in the mid-1600s was thought to have a special significance, with

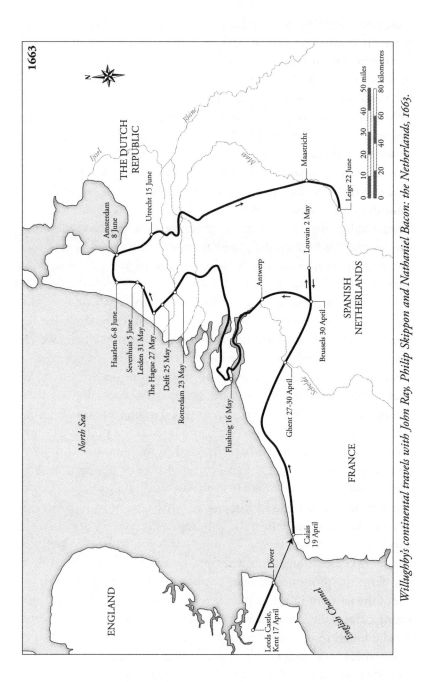

Willughby's continental travels with John Ray, Philip Skippon and Nathaniel Bacon: the Netherlands, 1663.

Descartes referring to it as the 'seat of the soul' – the vital link between mind and body.[32] The gland rests upon the butterfly-shaped sphenoid bone in the centre of the cranium, and it may have been this that Van Horne showed them.

On 5 June, before leaving Leiden, the party 'made a bye-journey to Sevenhuis [known today as Zevenhuizen], a village about four leagues [roughly five kilometres] distant, to see a remarkable grove, where, in time of year, several sorts of wild-fowl build and breed'. Travelling by boat they reached 'a most pleasant wood' occupied by a mixed colony of cormorants (which Ray refers to as shags), spoonbills, 'Quacks' and 'Regers' or herons. This must have been a remarkable experience. On their journeys through Britain, they had seen seabird colonies – albeit at a distance – but here they were in the heart of the colony. Under the nesting trees they were in danger of being splattered by the bird excrement showering down through the vegetation. Then there was the incessant and unavoidable screeching, belching and vomiting of the adult and immature birds above, and finally, the smell. To my mind and nose, the odour of heron colonies is curiously reminiscent of baked bread generously mixed with decaying fish and decomposing chicks that have fallen from their nest to die on the woodland floor.

They were astounded by the sight of cormorants nesting in trees and Ray considered them the only 'whole-footed birds' (birds with webbed feet) that are 'wont to sit in trees, much less build its nest upon them'. Their visit coincided with the annual harvest of birds, presumably mainly for human consumption, although according to another visitor, Gottfried Hegenitius, thirty years earlier, two boat-loads of live birds were sent to England each year. The method of harvesting, as related in the *Ornithology*, was that once the young herons, spoonbills and cormorants were deemed fat enough, 'those that farm the grove use an iron hook on a long pole to shake them out of the nest onto the ground'. Willughby was thus able to dissect a young cormorant, commenting on the blackish worms – parasites – that he found in its stomach. Ray must also have dissected a cormorant, probably elsewhere, for in the *Ornithology* he says that he too found worms similar to earthworms in it, along with an entire cod.[33]

Sevenhuis bird reserve in the Netherlands with its abundance of herons, spoonbills and cormorants – an undated drawing by Simon Klamputs, but probably 1700s.

It was at Sevenhuis that they also obtained a well-grown spoonbill chick, noting that it weighed 45½ ounces and measured thirty-four inches from the tip of its beak to the end of its toes; 'The colour of the body was snow-white like Swans.' This must have been the only spoonbill they ever got to see in the hand, for in the *Ornithology* their description is based on this same immature individual shaken from its nest. In Willughby's meticulous manner there is then an account of the bird's external and internal features. Externally, of course, its most notable feature was the unusually shaped bill – whose spatulate tip we now know is full of sensory nerve endings allowing the spoonbill to distinguish edible from inedible items as it sifts its beak from side to side through the water.[34] Internally, they noted that 'We did not observe in our bird those reflections [twists and turns] of the wind-pipe, which Aldrovandi mentions, describes and figures.'

Continuing, Willughby and Ray recorded: 'It has a large gall: the guts had many revolutions. Above the stomach the gullet was dilated into a bag.' Finally, they comment on the spoonbill's eggs, which are 'of the bigness of hens eggs, white and powdered with a few sanguine or pale red spots'.[35]

$$\vdash \curlyvee \dashv$$

Before they left England on their travels, Willughby had written from Middleton to Ray in Cambridge, telling him that he was going to design a cabinet for a collection of seeds. The exact date of that letter – now lost – is not known, but we have Ray's enthusiastic reply. It's a long quote, but it is important. He writes:

> You mention a *Box* which you intend for all *sorts of fruits and seeds*. It must have almost infinite Cells and Divisions to contain all the varieties of Seeds and Fruits. Concerning the Order and Method of it you need not my Advice, for I can give you none but what is very obvious, *viz.* to put those of the same Tribe near together. As for Instance, to have a *Drawer* with several Cells or Boxes for *Nuts*, another for *Cones, &c.* for the rest of *Fruits* which may be reduced to several Heads; and then one for *Exoticks*, which cannot be conveniently referred. In like manner

Aldrovandi's drawing of a dissected spoonbill showing the convoluted windpipe or trachea.

for *Herbs*, to have a Drawer with several Boxes or Divisions for *Legumina*, another the like for *Cerealia*, &c. only those Boxes must be more numerous than those for Fruits. By a *Drawer* with several *Boxes*, I mean such a thing as the Printers put their Letters in.[36]

The fact that Willughby was collecting seeds, and on a serious scale, belies the fact that when they decided to overhaul the study of natural history, they would divide the work into mutually exclusive zones such that Willughby would look after the animals and Ray the 'vegetables'. As this and much else makes clear, both men were involved in the study of plants and animals.

Willughby's cabinet still exists in the family home. This beautiful piece of walnut furniture consists of fifteen drawers elaborately divided by curved or linear metal strips into numerous compartments – of a different design in each drawer. In total, there are 1,200 compartments, and miraculously, given the passage of time, most of them still contain seeds, and there has been very little spillage. This is very surprising given that the cabinet has been moved between homes on several occasions, including in 1687 when Willughby's son Francis and daughter Cassandra went to retrieve their father's belongings long after his death. Twelve of the drawers contain seeds, and when I visited the Middleton family in 2007, Lady Middleton (now deceased) was keen to show them to me. As she pulled open a drawer I was surprised by the seeds' excellent state of preservation, but somewhat alarmed by the way they jumped around in their low-walled compartments as the drawers were shakily withdrawn and replaced. And, to my shame, I wasn't really interested, for at that time I had little sense of the significance of what I was being shown.

I have since wondered whether any of those seeds are still viable. They may well be because, as I learned from Richard Mabey's *Cabaret of Plants*, seeds as old as 2,000 years have germinated successfully. The other point Mabey makes is whether seeds lying inert for 350 years, like those in Willughby's cabinet, if still viable, can be considered 'alive'.[37]

Only when I went back in 2014 to photograph the seeds for a botanical colleague did I see the cabinet in a new light. Lord Middleton, Lady Middleton's son, met me and kindly agreed to help remove the drawers and place them on the floor so I could photograph them. He suggested that we start at the bottom and work our way up through the fifteen drawers. In fact, at first sight the two lowest drawers looked to be part of another piece of furniture for they lie below a small plinth; they appeared to have been constructed from a paler wood and, in contrast to all but the very top drawer, to have a centrally placed escutcheon. In fact, those two bottom drawers were part of the same piece of furniture; the difference in the colour of the wood was due to the middle twelve drawers being positioned behind two doors and hence protected from the light.

As Lord Middleton gently pulled open the lowest drawer and I saw what was inside, I gasped in disbelief, for here instead of seeds were birds' eggs. I was stunned. My mind raced. Could these be eggs that Francis Willughby himself had collected? If so, this would be among the oldest collection of eggs anywhere. Time was pressing: there were all those other drawers to photograph and Lord Middleton, I felt, was slightly bemused by my enthusiasm for what was essentially a tray of very dirty eggshells, many of which were broken.

We pushed on. I photographed the tray of eggs, we replaced it, and Lord Middleton opened the second drawer. Again, no seeds, but instead a bewildering array of mainly zoological curiosities. Scanning it quickly I noted some mollusc shells, but also fragments of large tropical beetles, some brightly coloured minerals, the dried skin of a skink, a beautiful scorpion and a spider made from wire, and a few detached labels. One compartment that made me blink in disbelief contained a collection of false human eyes. The skink made me wonder. Recognised since ancient times as a potent aphrodisiac, the tail of this particular specimen is missing. Had Willughby, or (given subsequent events) John Ray, been tempted to test its amatory effects?[38]

The layer of grime that covered everything seemed exactly right for something that was about 350 years old. Feeling as though I'd stumbled into an Aladdin's cave, I had a growing sense of the significance of what lay before me.

Working upwards, the next twelve drawers all contained seeds. I photographed them, noting that in each drawer was a folded sheet of paper comprising a map of that drawer's compartments and what they contained. The writing was obviously old, but also in several different hands. In one drawer was a vertically elongate, handmade book that was a catalogue of the drawer's contents. The handwriting here looked old too, but on opening its pages the paper was of a startling whiteness, giving the impression that it might have been made yesterday.

We came finally to the top drawer. I might easily have overlooked it since, like the lower two drawers, the wood was of a paler colour and appeared – at first sight anyway – not to be part of the cabinet's structure. It was locked and there was great consternation since no one seemed to know where to find the key. Lady Middleton went to look and although we tried a succession of different keys, we seemed to be out of luck. One by one the staff were summoned, but again no one knew the whereabouts of the key. Finally, someone who looked from his attire to be a gardener arrived and told us unhesitatingly that the key lay in one of the bottom drawers – and so it did.

This top drawer had no compartments and contained more mollusc shells, a few fossils, a piece of coral, a sea-urchin 'skeleton', some almost spherical stones, and fragments of crab exoskeletons. Most significantly, there were slips of paper and written labels on some of the shells that might provide a clue to the collectors.

The cabinet, so it seemed, was Francis Willughby's cabinet of curiosities, or part of it at least. Confirmation came when we examined the eggs in more detail. Most of them had been written on in brown ink. Francis Willughby had a distinctive 'hand' with certain letters, notably the 'p' produced in a particular way. It was a snipe's egg that clinched it, with the tail of the 'p' displaying his wonderful backward and forward sweep of the quill. Here was clear evidence that Willughby had labelled some of these eggs himself.

During my research career I have had a few Eureka moments, but this was one of the best. I simply could not believe that no one else – apparently – had seen or commented on these eggs or the other drawers of zoological specimens. Neither could I believe that so many of the eggs had survived intact. In fact, their survival was a miracle of neglect. Had they been displayed on cotton wool as is usual, rather than on the wooden base of the drawer, they simply would not have survived. This is because over time acids in the cotton fibres interact with the calcium of the eggshells so that they come to look as though they have been liberally dusted with caster sugar, an effect known as Byne's disease and one that eventually destroys the eggshells. Not a disease at all, as was once thought, this is a chemical reaction, in which the calcium breaks free from the shell. Had Willughby's eggs been displayed in this way, the drawer would have contained little more than dusty cotton wool.

The most miraculous thing of all was that Francis Willughby himself had written on some of the eggs. This provided unequivocal evidence that they were his.

As we chatted, Lady Middleton hinted to me that there had once been a second cabinet of eggs. That made sense because the majority of eggs in that lowest drawer were those of small birds, with nothing bigger than a heron's egg (which is only a little larger than a hen's egg).

The labelling on the several heron's eggs showed that they had been collected on 5 June 1663 when Willughby, Skippon, Ray and Bacon were at Sevenhuis in the Netherlands. The writing on the eggs, however, is not Willughby's, but it could be Skippon's, not least because the words on the eggs are identical to those in his journal for that day. I was confused initially because the bird's names on the eggs were either *Ardea cinerea*, which is the scientific name, both then and now, of the grey heron, or *Ardea cinerea minor*, which simply didn't make sense to me: what was 'minor', a smaller grey heron?

On looking more carefully, one egg has written on it the words 'Ardea major cinerea Common Herne Reyger Belg'. 'Herne' is the old English name and 'Reyger' (now 'Reiger') the Dutch for grey heron, and Belg means 'Dutch'. Another egg has the words 'Ardea cinerea minor Quacke Belgi', with 'Quacke' (now 'Kwak') the Dutch for the night heron.

The night heron would have been new to Willughby since it does not (and probably did not then) breed in Britain.[39] Confirmation that 'Ardea cinerea minor' refers to this species is clear from the *Ornithology*, where on plate 49 that name appears alongside an image that is unmistakably a night heron, albeit with the name 'night-raven'. This is becoming a bit convoluted, but 'Quacke' probably refers to the call made by this bird, which is active at night, and of which John Ray writes: 'It is called night-raven, because in the night time it cries with an uncouth voice, like one that were straining to vomit.'[40]

In total the egg drawer has twenty-eight rectangular compartments of different sizes, containing around 133 different eggs. It is impossible to be more precise because many eggs are broken or are mere fragments. Most eggs have been blown rather crudely through a hole at each end. It wasn't until the nineteenth century that collectors started to use an egg-drill, a tool rather like a countersink or pointed burr, which when twirled between the finger and thumb can create a single hole on the side of the egg through which the contents can be removed. Apart from the eggs of two heron species, most of those in the collection are of small songbirds that would have bred in the vicinity of Middleton Hall, including the yellow-hammer, chaffinch, house sparrow, starling, bullfinch, the pied, yellow and grey wagtail, wren and blackbird, but also the rook, green woodpecker and wryneck. This last, once widespread in England, is now effectively extinct as a breeding species here.

At least one mystery remains. Despite Willughby having assembled this collection at Middleton, eggs barely feature in the *Ornithology*. For many of the species – including the grey heron whose eggs lie in the seed cabinet – there is no description in the text of the *Ornithology*. Worse, in the account of the night heron, Ray (I presume it is him writing) says 'this bird lays white eggs' – which is incorrect and the specimen in the seed cabinet is the typical sky-blue egg of this species.[41]

⌇⌇⌇

Back to Sevenhuis. Dissection and descriptive anatomy were part of the innovative new science that was the driving force behind Willughby and Ray's cross-examination of the natural world. What *does* the inside of this bird look like? What does its internal structure tell us about its relations with other birds? Does the adult spoonbill's convoluted windpipe – as described by Aldrovandi – develop only as the bird matures? The answer to this last question is yes, as nicely shown by a study conducted in the 1970s.[42]

It is clear from the *Ornithology* that both Willughby and Ray made dozens of dissections, either together or separately. It is also apparent that Ray continued to dissect birds long after Willughby's death in 1672.[43]

At some point – we don't know when – Ray dissected a male green woodpecker, describing its reproductive anatomy and making an intriguing observation that I believe has neither previously been recorded nor followed up. Ray noticed that 'The right testicle is round, the left oblong, and bent almost into a circle, which lest anyone should think accidental, I observed in three several [separate] birds.'[44] This strongly suggests that the odd-shaped left testicle is a standard feature in this species. Over my career I have dissected hundreds of birds (killed by cars) to examine their reproductive anatomy, but not a male green woodpecker. In many bird species the two testes differ in shape and size, but I have never seen anything as extreme as that which John Ray describes. Intriguing!

After reporting on the green woodpecker's organs of generation, Ray switches his and our attention to the tongue, which 'when stretched out is of a very great length, ending in a sharp, bony substance, rough underneath, wherewith, as with a dart, it strikes insects'. He goes on to say – and this suggests he watched a live bird – that the tongue can dart out three or four inches and 'draw up again, by the help of two small round cartilages [the hyoid bones], flattened into the fore-mentioned bony tip, and running along the length of the tongue'. Ray goes on to describe how the cartilages 'from the root of the tongue take a circuit beyond the ears, and being reflected backwards towards the crown of the head … make a large bow'. This – and there's more of it – is a very careful description of what is an incredibly demanding dissection. Neither Ray nor Willughby was the discoverer of the bizarre arrangement of the woodpecker's and wryneck's tongue – Leonardo da Vinci had reported them previously – but this gives a good idea of the level at which Ray and Willughby were operating (literally) and the fact that they needed to see such curiosities for themselves.

No fewer than 155 of the 380 birds described in the *Ornithology* include an account of their internal anatomy. Willughby's propensity for dissection is reflected in a letter of May 1672 from Henry Oldenburg to Martin Lister, in which Oldenburg refers to Willughby's practice of dissecting birds, fishes and quadrupeds – and finding worms in most of them, just as he had with the young cormorant at Sevenhuis.[45]

Picus viridis
The common green Woodpecker
or Woodspite.

Caput Pici dijsectum.

Green woodpecker (or woodspite), from the Ornithology of Francis Willughby *(1678), showing the curious arrangement of the tongue.*

From Leiden, Willughby and company travelled north to Haarlem and Amsterdam, and from there they rode south to Utrecht, spending a few days at each and arriving at Maastricht on the evening of 20 June 1663. In his journal Ray commented on the white storks nesting on the 'chimnies in the towns and cities as well as villages'. He also mentions the constant chiming of bells, 'which seldom rest and were to us troublesome with their frequent jangling'. He could not help but notice the extreme cleanliness of the houses (presumably compared with England), with all 'house-hold stuff marvellously clean, bright and handsomely kept: nay, some are so extraordinarily curious, as to take down the very tiles ... and clean them.'[46]

In his account of their journey Ray also commented on a difference between the women of England and the Netherlands, quoting from his deceased Trinity College friend, Sir Francis Barnham, who had been there in the early 1600s:

The common sort of women (not to say all) seem more fond of and delighted with lascivious and obscene talk than either the

English or the French. The women are said not much to regard chastity whilst unmarried, but when once married none more chaste and true to their husbands. The women even of the better sort do upon little acquaintance easily admit saluting with a kiss; and it is familiarly used among themselves either in frolicks or upon departures and returns.[47]

Not surprisingly, food, especially at inns, was an important topic of 'research'. The Dutch, they noticed, were almost always eating. The first dish of a meal is usually salad, 'Sla they call it, of which they eat abundance ... The meat they commonly stew, and make hotchpots of it. Puddings ... they do not eat; either not knowing the goodness of the dish, or not having the skill to make them.' Judging from a letter that Ray later wrote to the physician Martin Lister about his diet (and poor health), Ray was inordinately fond of puddings and other rich food – presumably a taste he acquired in Cambridge.[48]

The common people, Ray says, consume cod and pickled herrings 'which they know how to cure or prepare better than we do in England'. The excellent smoked beef 'they cut into thin slices and eat with bread and butter, laying the slices on the butter'. They have four or five sorts of cheese, including 'those great round cheeses colour'd red on the outside, commonly in England called Holland-cheeses'. Their beer, 'thick beer they call it and well they may', is more than three pence a quart. Finally, 'all manner of victuals, both meat and drink, are very dear'. Not only that, every transaction with an inn-keeper, as well as with waggoners, porters and boatmen, involved bargaining, not always very successfully since 'the inn-keepers, in many places, exact according to the rich habit and quality of their guests'.[49]

Such was the nature of continental travel for the wealthy.

5

Images of Central Europe

By late July 1663 Willughby's party had reached Strasbourg in the Alsace. They were there for just three days, but it was here that Francis learned of a local man whose life-long hobby was to have an important impact on his and Ray's natural-history makeover. Leaving his three colleagues in the city, Willughby went off on his own to meet Leonard Baldner, a dapper little man in his early fifties sporting a reddish handle-bar moustache and chin-puff beard. A prosperous, well-educated member of the city council, Baldner served as the official keeper of forests and Strasbourg's three rivers: the Rhine, Ill and Breusch, where he was permitted to hunt and fish as he wished.

Proudly, he told Willughby how he had been born in Strasbourg in 1612, and how he now (after three marriages) had a brood of twelve children. In 1646, he said, when in his thirties, he had shot some beautiful waterfowl he didn't recognise. After arranging for someone to paint them, he was so struck by the result he decided to describe and illustrate all the birds, fish, quadrupeds and invertebrates that he subsequently killed or captured. The result, as Willughby could see, was a comprehensive, exquisitely illustrated (unpublished) guide to Strasbourg's aquatic birds and other wildlife.

Baldner spoke and wrote with enthusiasm and originality and in way that Willughby had probably not encountered previously. Here was someone driven by curiosity, very much like himself, but in Baldner's case, fuelled by a unique connection with the watery world in which he worked. Willughby was thrilled.[1]

We don't know who tipped Willughby off about Baldner, but his colleagues apparently didn't think it worthwhile to accompany him that day. One can only imagine Willughby's delight at meeting the man and – possibly with a guide and interpreter, or just possibly with Philip Skippon who almost certainly spoke German – hearing about his passion, and then seeing his extraordinary and beautifully illustrated homemade book. In a world where published zoological illustrations consisted of black-and-white engravings or crude woodcuts, the vibrant colours and lifelike poses of Baldner's animals must have seemed incredible. Willughby decided there and then to buy the book. Measuring just 30 by 20 centimetres and made up of some 156 pages, it comprised both text (in German) and paintings of fifty-six birds and forty fishes as well as various mammals, amphibians and an array of aquatic invertebrates.

Regardless of how detailed a written description was, Willughby knew that there was no substitute for a superbly executed, coloured image to convey exactly what a particular bird, fish, dragonfly or dormouse was really like. Good pictures also inspired one's readers. Baldner's book is a gem.

⊢⋏⊦

Willughby's copy of Baldner now lies in the British Library in London. It is dated 1653, which is presumably when it was completed, and is accompanied in the library by a translation of Baldner's German text made after Willughby's death. In his volume on John Ray, Charles Raven refers to this book, rather condescendingly it seems to me, as Baldner's 'not very literary notes'.[2] Let's look at what Willughby would have learned. Baldner records where and when he shot the birds and whether they were worth eating – both of which would have been of limited interest to Willughby, since these topics were not the focus of his and Ray's studies. Baldner's comments on the birds' ecology and behaviour, however, would have been very germane. Of the kingfisher, for example, which Baldner noted 'smells very rankly' and is not good eating, he said that it makes its nest in a deep hole in the riverbank 'exactly as straight as a carpenter's square' and the female lays her eggs in a small chamber at the end, fledging five or six young in August. 'And when one findeth a nest and doubteth whether or not there be young ones therein let him observe in the morning and he shall hear the young ones crying, or when their dung runneth out of the hole, then be sure there are young ones therein.'[3]

To prepare birds for the table Baldner had to open them up, but being of an inquisitive nature, this was done more as dissection than butchery. Indeed, Baldner's knowledge of his victims' internal anatomy may have made his work especially attractive to Willughby. Of the bittern (again, not 'dainty eating'), Baldner commented on its huge intestines that he felt must in some way enhance their extraordinary booming courtship call – a call that sounds like someone blowing over the top of a large bottle. In Baldner's day most people believed that bitterns boomed either by blowing through a reed or by thrusting their beak into the mud, but Baldner watched booming bitterns and saw for himself that the sound emerged through their closed bill 'lifted high up'. Willughby and Ray mention the bittern's unusual call in the *Ornithology*, but they offer no explanation for how it is made, other than the erroneous folklore. Had they read Baldner's account they would have known the truth.

The stone curlew is a rare bird in Alsace (which lies just outside its current breeding range in Western Europe), and Baldner only ever saw one. He shot it, and recognising that it was unusual, made sure he produced a painting of it highlighting one of its most unusual features – a characteristic mark – the lack of a back toe.

Discussing the common wild duck or mallard, Baldner makes an intriguing observation that 'they have a very quick scent, in so much that they smell out a man though they do not see him if only they have but the wind of him'. While it is often difficult to ascertain which of its senses a bird has used to detect a predator such as man, there is now good evidence that the sense of smell in birds – once thought to be all but non-existent – is well developed in certain species, including the mallard.[4]

Baldner also kept some animals in captivity, including pet otters; a gull he had hatched in a homemade incubator;[5] and a tame cormorant – a scarce bird on the Rhine – that was 'tied to a string, who catch'd the fishes himself that he eats'.[6] It is widely known that cormorant fishing was (since the third century) common practice in China, but it is less well known that European cormorants were also used in this way. In the *Ornithology* Ray gives a second-hand account from Johannes Faber, who describes how captive cormorants are 'hood-winked' (i.e. have a hood placed over their eyes, as falconers do with their birds), 'that they be not frightened', and 'When they [their keepers] come to the rivers they take off their hoods, and having tied a leather thong round the lower parts of their necks that they may not swallow down the fish they catch, they throw them into the river.'[7] Once the cormorants have caught five or six fish the birds are called to their keeper, and 'little by little one after another they vomit up all their fish a little bruised with the nip they gave them.'[8] James I of England employed a keeper of cormorants in the early 1600s and enjoyed watching his captive birds fishing. So popular was this aquatic form of falconry that the king had birds imported both from the Isle of Man and Reedham in Norfolk, but also from Sevenhuis in the Netherlands where he ordered 'yearly two ships full'. The cormorant fishing tradition was continued by Charles II, who instructed his servant Richard Edes to 'keepe and breede three cormorants for our recreation'.[9]

The illustrations of the fish in Baldner's book are arguably more beautiful than the birds, and he clearly liked fish and knew a great deal about their biology. He describes a wels, *Silurus glanis* (the sheat-fish or European catfish), kept in a pond for five years, a fish that reached five feet in length. This is Europe's river monster. They can grow even bigger though: the record is 2.8 metres (9 feet: 144kg or 317lb); it is an ugly pin-eyed fish, with an enormous head and a cavernous mouth. Baldner's image of it gives no sense of its size and gluttonous ferocity, although I like the fact that his specimen does have an evil glint in its tiny eye.

Baldner knew all about the curious reproductive and anadromous (migrating up rivers from the sea to spawn) habits of salmon – facts that took others two more centuries to rediscover. His relentless curiosity is a joy to observe: just how many eggs has this gravid female pike got inside her? He counts them and finds an astonishing total of 148,800.[10]

I was somewhat puzzled to see mentioned in the Preface to the *Ornithology* that Baldner's notes were appreciated, with Ray commenting that his curiosity was 'much to be admired and commended in a person of his condition and education ... For my part, I must needs acknowledge that I have received much light and information from the work of this poor man, and have been thereby inabled [sic] to clear many difficulties, and rectify some mistakes in Gessner.' This is puzzling given how little of Baldner's information appears in the *Ornithology*.[11]

We don't know how much Willughby paid for his copy of Baldner's book, but it seems likely that Baldner was thrilled, both with the money but also by the fact that someone of Willughby's standing considered it worth purchasing. It is possible that Baldner anticipated something like this happening for there are several – at least six – copies of his handcrafted book in existence. He cannot have sold his sole copy to Willughby and then decided to create another, for the illustrations and text are similar across the various versions. How and why Baldner produced multiple copies is not known.

The copy in the British Library is annotated in Willughby's hand, and science historians are confident that this was the one bought by him. We do not, however, know how it ended up in the British Library.

For a long time only three other copies of Baldner's book were known to exist: two in the Strasbourg Library, one of which was destroyed by a fire in 1870, the other thought to have been (very poorly) illustrated by Baldner's twelve-year-old son, Andreas. The third copy, inherited in 1686 by Landgraf Karl von Hessen, later placed in the Heidelberg Library, and subsequently in the library at Cassel, is generally considered to be the most beautiful. It was studied by Robert Lauterborn in the early 1900s and a very attractive, but expensive, facsimile was made of it in the 1970s.[12]

In the 1920s another, previously unknown, copy emerged, and luckily its owner – John C. Phillips, who obtained it through an English dealer – decided to publish a short account of it. With fifty-seven bird plates, forty of fish, three of mammals, and seven of reptiles, amphibians and invertebrates, it was similar in many respects to Willughby's copy in the British Library. One difference, however, was the inclusion of 'a large collection of inferior pictures with French and German legends' – very obviously not by Baldner, but curiously, containing a portrait of him.

Enchanted by his acquisition, Phillips wrote: 'Some of the smaller species of insects and their larvae are so delicately drawn and colored that one looks at them with astonishment and admiration; they are so much better than average pictures of that period'.[13] Indeed!

In the mid-1930s Phillips's copy of Baldner passed into the hands of Albert E. Lownes, owner of one of the most important collections of books on the history of science in the United States.[14] On examining his newly acquired copy of Baldner, Lownes noticed a number of intriguing things. First, that it bore the same date – 1653 – as the London copy, whereas the other known copies were all dated 1666. He also commented that rather than being a haphazard assortment of images, this was a carefully assembled collection; that the illustrations were by several different artists; that they included some preparatory sketches in pencil, ink or sanguine (a reddish-brown, iron-oxide chalk, sometimes mixed with water to form ink), which Lownes suggests would have been preserved only by a working naturalist (rather than a collector); that many illustrations were labelled with locations, including Augsberg, Antwerp, Brussels and Amsterdam, or annotated with phrases such as 'on the way from Florence to Siena', and that many of the images carried the name of earlier naturalists such as Belon, Aldrovandi, Marcgraf, Clusius and others. Even as I write this, reading from Lownes's report I can feel his – and my own – heart rate rising. At this point he must have begun to realise what he had before him. The final piece of the jigsaw was provided by a drawing labelled 'Picturam transmisit Th: Browne, MD' – a picture from Thomas Browne. As was known, Browne helped John Ray as he prepared the *Ornithology* by providing him with a number of excellent bird images.[15]

Putting two and two together, Lownes realised that Francis Willughby must also have owned this copy of Baldner. He speculates that Baldner might have presented the second copy to Ray, but this seems unlikely for there's no mention of it in Ray's journal. Another possibility is that having seen Willughby's copy, Skippon purchased the second one, but neither does he mention Baldner in his travel diary.

The other thing that is curious about all this is that in the preface to the *Ornithology*, Ray talks about Willughby purchasing 'a' copy of Baldner. A puzzle still waiting to be resolved.

Several accounts of the Baldner volumes assume that they were illustrated by him. One copy, now in private hands and sold at Christie's auction house in 1995 for £87,300, may well be the original since it contains some 'less accomplished' watercolours that could be by Baldner himself. In contrast, the two Willughby copies and the Cassell one were illustrated by a much more accomplished artist – Johann Walther. That name is potentially confusing, for three different Johann Walthers – a father and two sons – are all associated with Baldner. The father is Johann Jakob Walther, a very skilled bird artist who enjoyed royal patronage and whose consummate images can be found (if not easily viewed) in the Albertina Museum in Vienna. His two sons were Johann Georg, born in 1634, and Johann Friedrich, born in 1639. It was Johann Georg, apparently a distant relative of Baldner's,[16] who executed the illustrations. As an artist he was good, but not as good as his father, with whom he shares some similarities in style and manner of labelling his artwork.[17]

†↑†

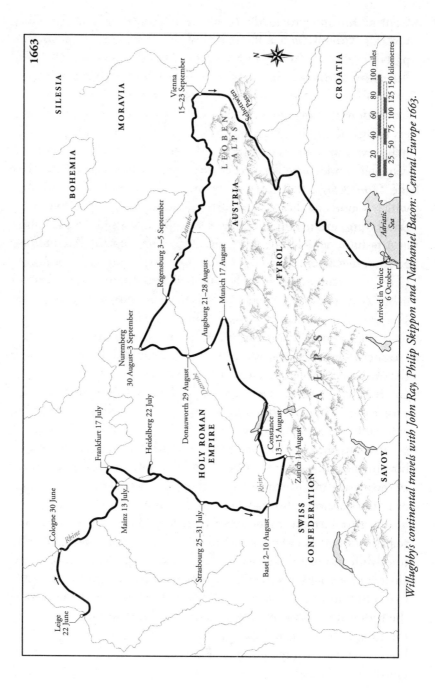

Willughby's continental travels with John Ray, Philip Skippon and Nathaniel Bacon: Central Europe 1663.

Central Europe provided other ornithological riches. Travelling towards Nuremberg, via Basel, Zurich and Constance, Willughby's party shot a vivid blue, crow-sized bird on 21 August 1663 near Augsburg: a roller. It may even be the bird whose rather crude image, as a corpse, lies (or rather, hangs by its beak) in the Middleton Collection at Nottingham. Although Ray and Skippon saw rollers subsequently in Sicily and Malta, the Augsburg specimen – a male – was the first they had seen, the one Willughby measured and dissected, and whose detailed description subsequently appeared in the *Ornithology*. It is referred to there as the 'Strasburgh Roller', which sounds as though Ray might have confused Augsburg and Strasbourg, even though Skippon[18] clearly says the bird was killed at Augsburg, which he and Ray both agree was where they were on 21 August. However, it seems that Ray – albeit with no explanation – was using a name employed by some previous writer.[19] The great Victorian ornithologist Alfred Newton somewhat cryptically suggests that this was Conrad Gessner (1555), but Gessner never used that name and instead refers to the roller as 'the blue crow'. Gessner's knowledge of the roller – and that name, derived from the bird's aerial rolling display – came from his friend Lucas Schan, an artist who sent Gessner paintings and skins of the roller. Schan wasn't the only artist to paint the roller. Albrecht Dürer made several paintings of the wing of a roller between 1512 and 1524 in what are perhaps some of the most iconic and beautiful of all medieval bird images.[20] It is just possible that Willughby may have seen one of these paintings since we know that he and his friends saw some of Dürer's artwork during their travels.[21]

It wasn't just the roller's rich blue, black and purple plumage that impressed Willughby and Ray; they also noted four very distinctive features, or 'characteristic marks'. These were, first, a tiny cluster of wart-like excrescences behind the eye; second, that the two outermost tail feathers were longer than the others; third, that the toes were completely seperate from their base; and finally that the tongue had two forked appendices. I checked with some colleagues who are studying rollers and although they knew about the tail feathers, they hadn't noticed the three other features, perhaps because if you are an ecologist it isn't usual to examine a bird's

tongue, toes or eyes! In terms of the wart-like excrescences behind
the eye, I was able to see this by zooming in on some pictures
on the Internet. The toes were a bit more of a challenge, but in
many birds some degree of fusion between the toes is widespread,
and indeed in describing the bee-eater,[22] a relative of the roller,
Willughby comments that the 'fore-toes … are all joined together
to the first joint, as if they were but one toe'. This is a state referred
to as syndactyly, both in birds and people (where it is a rare genetic
disorder). As a taxonomic feature among birds it isn't apparently
very useful, but in the 1660s Willughby didn't know that, and his
comment on the absence of syndactyly in the roller is further evidence of his meticulous and knowledgeable approach.[23]

Nonetheless, I was left rather perplexed by those four distinguishing marks because it felt a little like – to mix zoological and
botanical aphorisms – not being able to see the wood for the trees.
How many cerulean, cobalt and chestnut crow-sized birds had
Willughby encountered? The dazzling plumage is the single most
important distinguishing feature of this species.

On the other hand, had Willughby known that there are (today)
eight recognised species of roller in the world, finding some features that distinguished them would be (and has been) important.
Perhaps his focus on the minutiae of the European roller was a
symptom of the over-emphasis on detail for which John Ray gently
chided him. Even so, those four features that Willughby noted are
of interest in their own right, and as I have indicated, some of them
may in fact be characteristic marks that distinguish rollers from
their closest relatives such as the bee-eaters.

At Nuremberg, which the group reached on 28 August, they
saw that 'birds alive of all sorts are brought everyday into the
market and they sell (to eat) jays, starlings, wrens, titmice &c'.
While the idea of eating wrens or titmice might seem abhor-
rent to us today, that's only because the creatures have been
sentimentalised by us. In seventeenth-century central and southern Europe – and indeed in certain regions still – small birds
were simply there alongside fish, crustaceans and snails as part
of the diet. However, the fact that Skippon felt it was worth

commenting on suggests that he was surprised by it and that such practices were uncommon in England.[24]

Birds commonly eaten in England in Francis Willughby's day

bayning – unidentified	pewits – black-headed gull
bitterns	pheasants
buntings	pigeons – woodpigeon and other doves
bustards – great bustard	plovers – probably golden plover
char – Thomas Browne's 'churr', a small wader	puffins – Manx shearwater
cranes	quail
curlew	redshanks
dotterel	rils – unidentified
fieldfares	roe – unidentified
godwits	sand[er]lings
heath-cocks – black grouse	sea plover – grey plover
herons	snipe
knot	stint
lapwings	swans – mute swans were eaten at feasts; may include whooper and Bewick's
larks – skylark	teal
maychit – mentioned by Thomas Browne: a small, very fat wader	throstles – song thrushes
merles – blackbird	thrushes – song thrushes
moor-pouts or grouse – red grouse	wheatear
ouzels or blackbirds – either ring ouzels or blackbirds	wild duck – mallard and presumably other species
partridge – grey partridge (the red-legged partridge was introduced later)	wild geese – presumably several species
	woodcock

Note: The list of birds is from Edward Chamberlayne's (1676) list of abundant English foods in *Angliae Notitia or the Present State of England.* Some identifications are from Swann (1913).

It was during their three days in Nuremberg that Willughby acquired another set of bird illustrations. We know this because Ray mentions it in the preface to the *Ornithology*: 'At Nurenberg [sic] in Germany he [Willughby] bought a large volume of pictures of birds drawn in colour.'[25] Curiously, neither Ray nor Skippon mention the acquisition of these pictures in their travel diaries. It is possible that Willughby went off on his own again to negotiate a price, but it is hard to imagine the other members of his party being unaware or indifferent to the paintings. Nor do we know from whom Willughby purchased the pictures, or who executed them. They still exist, at least in part, in the Middleton Collection, and judging from the diversity of styles and quality, several different artists were involved. Some of the illustrations like the cuckoo are superb, whereas others, such as the small passerines perched on branches decorated with unconvincing sprays of stunted leaves, are both unrealistic and unattractive.

This collection of images was transferred from Willughby's descendants, the Middleton family, to Nottingham University Library in 1947. The pictures have been scrutinised by several historians and perhaps most intensively in the early 2000s by Nick Grindle, now at University College London. I have looked at these images too and they comprise a mixed and – to me at least – a visually rather disappointing collection. Grindle's ability to find some order there is impressive. He suggests that up to nine different artists were involved, and notes that many of the 116 bird images are annotated, some by Willughby himself and always in Latin, as was his way, but also perhaps by their previous owners in Dutch, French or German. Of the fish, there are some eighty-two paintings or drawings by twenty-four different artists and another group of fourteen bearing the signatures of Willughby's colleagues, including Philip Skippon, Nathaniel Bacon and Francis Jessop, but whether they executed them or merely lent them is unknown.

One tiny and easily overlooked clue as to how Willughby used his bird images is – as Grindle noted – that many of them have tiny pin holes in their corners. When this was first pointed out to me, it conjured up images of a historian with nothing else to go on desperately clutching at any clue that might help him or her interpret what they were seeing. As I became more familiar with the Middleton Collections, with the way historians work, and with Willughby himself, I found myself agreeing with Nick Grindle's interpretation. His suggestion is that the holes tell us the illustrations were pinned onto backing sheets so that the bird images could be viewed together. Just think about it for a minute: it is so easy today for us to consult a field guide to see how a rook, crow or jackdaw compare. In Willughby's day images were scarce, invariably monochrome and often poorly executed, making it difficult, to say the least, to see similarities and differences between species and perhaps identify natural groupings. Also, as Grindle suggests, the motivation for Willughby's visual comparisons was the overarching desire to create a classification.

Journeying by coach from Nuremberg, Willughby and his colleagues travelled through beautiful pinewoods to arrive on the afternoon of 3 September 1663 in Altdorf, a tiny, walled town, whose houses they noted were 'indifferently built'.[26] The university at Altdorf conferred degrees of 'doctors of law, physic and poetry, bachelors of divinity and matters of art',[27] but it was the university's physic garden that they had come to see. This was created by Moritz Hoffmann, professor of medicine and botany and the author of an impressive illustrated catalogue of medicinal plants,[28] who greeted his visitors with great charm. Knowing about his catalogue before they had set off from England, Willughby and Ray must have had the physic garden on their itinerary right from the start.

Hoffmann showed them two books of dried plants containing some 3,000 specimens: the first comprised specimens primarily from the Padua physic garden; the second were plants from the Altdorf garden. Willughby and his colleagues were also treated to several curiosities including the bones found in the human ear; 'a little wooden head curiously imitating all the futures [features] and other parts of a human head'; an eye of box (wood) and 'another of ivory, with the optic nerve, tunicae, humours &c'. They also marvelled at an early modern wheelchair designed by a lame person from Altdorf, enabling him to get to church 'without any help'. Philip Skippon was so impressed that he included a diagram of the wheelchair in his journal.[29]

Trained in Padua, Hoffmann was one of several physicians just starting to become interested in minerals and fossils, and he proudly showed his visitors his collection. These included 'pectinites' – scallop-like fossils, glossoptera, sea balls from Naples and ammonites. Glossoptera were so-called because of their tongue-like shape (*glosso* = Greek for 'tongue') and were once thought to be the petrified tongues of dragons or snakes. Their true identity – fossilised sharks' teeth – was revealed by the Italian philologist and antiquarian Fabio Colonna in 1610. Sea balls are mineral concretions rather than true fossils, which weather out of marine cliffs and whose spherical appearance encouraged their collection as curios.

As anyone with even the slightest interest in geology knows, ammonites are the commonest fossils, and in Willughby's day they were referred to as serpent stones or snake stones. The name arises, obviously, from their coiled, snake-like form, but also because these fossils are particularly abundant near Whitby on the Yorkshire coast, where in the seventh century St Hilda, Whitby's abbess, banished a plague of snakes by turning them to stone. That these particular snakes lacked a head was thanks to a beheading curse by Hilda's contemporary, the hermit of Lindisfarne, St Cuthbert, who, incidentally, had a tame eider duck that followed him everywhere.[30] So strong is the link between St Hilda and ammonites that one genus now bears the name *Hildoceras* and three ammonites adorn Whitby's coat of arms. The snake myth was reinforced and given credence in Victorian times by local artisans carving serpent heads onto the fossils.

Ammonites are the petrified remains of the shell of extinct cephalopods – close relatives of the octopus and cuttlefish. They were common in the seas 200 to 400 million years ago and became extinct about 65 million years ago. Their shell comprises a series of gas-filled chambers by which the animal could regulate its position in the water column, and its soft body – complete with tentacles and a beak – occupied the final outermost chamber. The structure of the animal itself is unknown because it never fossilised, but the shell's similarity to that of certain extant cephalopods allows palaeontologists to confidently infer what it was like.

No such confidence existed in Willughby's day. Fossils were a mystery and – as exemplified by serpent stones – they are often imbued with convoluted tales regarding their origin.

The day after visiting Hoffmann the group travelled on to Neumarkt, where as Skippon tells us, 'two miles [3 km] further we lodged this night in the straw at a poor village [whose name he had forgotten] where we found *Cornua Ammonis* [ammonites]'.[31] In his account, Ray says that in some fields they 'gather'd up [a] good store' of fossils for themselves. Some of those ammonites may well be the ones that still lie in the uppermost drawer of Willughby's cabinet.

In the mid-1600s fossils were a problem. What were they exactly? Sharks' teeth did indeed seem to be what Fabio Colonna said they were, as was obvious from a comparison with the teeth of living (or rather, recently dead) sharks. On encountering fossil stems of giant horsetails (*Equisetum*) in Italy the following year, John Ray immediately recognised their similarity to living horsetails, allowing him and Willughby to conclude that 'the parts not only of trees but also herbs themselves may sometimes petrifie'.[32] Ammonite fossils were an issue, however, because – like dinosaur fossils discovered later – they didn't obviously resemble any living organism. And if God had created all known organisms, what were these unknown things? Even more puzzling was the location of marine fossils, like cockles and scallops, far from the sea and at the tops of mountains.

In Bologna, Willughby and Ray visited Aldrovandi's museum and examined his collection of fossils. They were already familiar with his comprehensive and thoughtful views on marine animal remains described in his *De Reliquis Animalibus*, and in his posthumously published *Musaeum Metallicum*.[33] Both books contained plenty of images and ideas, including the notion that fossils, like crystals and other minerals, had 'grown' – a view that both Willughby and Ray could relate to as a result of their experiments with chemical gardens, in which lifelike but inanimate structures grew in front of their eyes.[34] Given their reliance on the great Italian naturalist's work, it is perhaps not unexpected that some of Willughby's fossils are labelled with reference to Aldrovandi's books.

Exactly at the time when Willughby and his friends were travelling across the continent, the question of what fossils were was being actively discussed at the Royal Society in London. Robert Hooke – extraordinarily brilliant, physically deformed and not universally popular – examined fragments of petrified wood using his newly

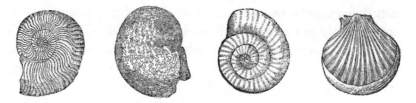

Examples of fossils (l–r): three different ammonites and a scallop-like bivalve, all from Aldrovandi (1648)

acquired microscope in 1663, confirming that the specimen's fine structure was exactly like that of extant trees. He concluded that fossils were the organic remains of long dead and – in some cases – extinct organisms. As though this wasn't clever enough, he also came up with the earth-shattering idea that marine fossils found on the tops of mountains had got there as a result of earthquakes. Hooke's ideas challenged conventional beliefs about the age of the earth and the biblical narrative, and the ensuing debate rumbled on for years.

Interestingly, Hooke's wonderful insights had been anticipated in part at least by Xenophanes in the sixth century BC, suggesting that the impressions of leaves and all manner of sea creatures embedded in rock were the result of everything being covered in mud long ago, after which the impressions dried out.[35]

When he reached home in 1666, Ray became aware of the Royal Society's fossil debate, and anxious to demonstrate how up to date he was with scientific developments, he decided to include a fourteen-page digression on fossils in his account of the continental journey, listing where they could be found and their probable origin. This was based partly on his own experience and partly on what he calls other 'good writers'.

Ray accepted the biological origin of fossils because to do otherwise would 'put a weapon into the atheists hands'. If fossils were sports of nature created 'without reason or function … then living animals could also be produced without counsel or design', forcing Ray to accept that fossils were once living creatures. But he wasn't happy. The problem was that some fossils – like ammonites – looked like nothing on earth and therefore, if they were once living organisms, they must be extinct. Ray's religious beliefs simply would not allow him to accept that God might have created certain animals or plants only to allow them to be subsequently lost.[36]

ﻮﻮﻮ

Willughby's company entered Vienna on 15 September 1663, and visiting St Stephen's Cathedral, looked in awe at the panorama of the city that was their reward for climbing the 414 stone steps to the top of the spire. In the narrow streets – busier, they said, than anywhere they had been other than London – there were Hungarian soldiers, proud, confident and curiously exotic. Dressed in boots and fur caps adorned with two or three long feathers and armed with a poleaxe and long, broad-bladed scimitars, they were an imposing sight. Most 'are habited all in blue … some of the better sort wear black. Many have their heads are shaven, except one lock, which they let grow on the top of their heads. We saw some of their gentlemen on horseback, with leopards skins wrapt about them …'[37] These were the victors of a pitched battle between an Austrian-Habsburg army and the Ottomans – the Battle of Saint Gotthard and part of the Fourth Austro-Turkish War of 1663–4 – that had occurred just six weeks earlier on the Raba river some eight kilometres south of Vienna.[38]

Visiting the markets, Willughby and his friends were fascinated to discover that they could buy land tortoises for about six pence apiece, 'which are good meat when their heads and feet are cut off; they are found in these parts in muddy ditches'.[39] To me these sound more like European pond terrapins, which are common around Vienna, whereas land tortoises occur only further south.

In a house in the Viennese suburbs they came across a picture of a *Hausen* fish – that is, a sturgeon – 'of a great bigness' that had been taken from the Danube 'frequently bought hither in Lent', adding that 'Of the spermatic vessels 'tis said the ichthyocolla is made.' Ichthyocolla refers to isinglass, the same substance Willughby and Ray used when growing their 'chymical' gardens in Cambridge. Skippon isn't quite correct in saying that the isinglass is made from the Sturgeon's 'spermatic ducts'; it is from the swim bladder. The sturgeon was the most popular fish for this purpose, probably because of its size, but the swim bladders of many different fish also yield the collagen-based substance from which isinglass can be prepared.[40]

Significantly, Vienna provided an unusual opportunity to add to their word lists, first tested on their Welsh trip the previous year. Now, however, Willughby's approach was to ask his informants for the Lord's Prayer, and his prompt list comprised Latin rather than English words: 'We busied ourselves with several persons in procuring Bohemian [Czech], Hungarian, Polonian [Polish] and Turkish words.'[41] Skippon transcribed the Turkish words on a prompt list of his own making, and made notes on both phonetics and spelling. Willughby and some unknown persons – judging from the writing – took care of the rest, adding them to a prompt-list scroll on which had already been listed samples of Low Dutch (Dutch) and High Dutch (German). This was a joint project between himself and Philip Skippon, who continued to collect information, including Romansh and Maltese words when he and Ray later separated from Willughby and Bacon in Italy.

After nine days, Willughby's party left Vienna on 24 September 1663, and headed south across the Alps to Venice and to a new world.

6

Italian Sophistication and Spanish Desolation

The journey from Vienna to Venice was tough. The weather was poor, the mountain tracks were steep, and throughout the entire sixteen-day journey Willughby and his friends had little to eat other than bread and 'what was made of sorghum'. Even so, they managed to enjoy the discovery of some new Alpine plants and birds including nutcrackers and 'mountain finches' (bramblings). They were shocked, however, by tiny communities in the mountain valleys near Leoben in which the men and women had 'great bronchoceles ... some of which were single, others double and treble' under their chins. Also known as 'Bavarian Pokes' these were goitres caused by enlarged thyroid glands. Worse, many of these people seemed to be 'ideots and scarce sound of mind'.[1]

Since Roman times it has been known to some at least that the people of remote Alpine valleys, once referred to as cretins, often suffered from goitres, mental deficiency and stunted growth. The cause of this condition, known now as congenital hypothyroidism, was originally attributed to stale mountain air. The true cause, however, discovered in 1851, was a lack of dietary iodine, an accident of geography and geology, exacerbated in this case by inbreeding.[2,3]

Descending from the mountains onto the sun-drenched, plant-rich plain near Friuli their spirits soared, with Ray noting that 'This part of Italy hath been deservedly celebrated for its fertility, and may justly, in my opinion, be styled the garden of Europe.'

It was about to get better. Venice – where they arrived on 6 October 1663 – proved to be an educational utopia. Rich in wonders, both natural and artificial, crammed with culture, both polite and impolite, Venice perched precariously on the watery cusp between the exotic east and the conservative west. Then as now, Venice felt like a fairy-tale city, with its winding waterways, endlessly confusing narrow streets, luminescent light and wonderful architecture. Little wonder that Willughby and friends spent three months here. They loved it: 'We never enjoyed our health better, nor had better stomachs to our meat in any place beyond the seas than we had here.'

Venice was also a biologist's paradise: the fish market was – and still is – better than the best zoology practical class, with over sixty species of fish and more than twenty marine invertebrates, many of which they were seeing for the first time. The market also (then, but not now) offered an extraordinary variety of dead birds for sale.

After their exhausting journeys, it must have seemed as though they were on holiday. There was so much to see and do, including the welcome distraction of theatrical performances. On one occasion, they went to a comedy 'where at the door we paid 16 soldi, when others paid but six'. Skippon describes the scene: 'Round about were four or five rows of boxes ... where Venetian gentlemen and others sat ... The gentlemen, and some with their wives or whores, came masked and disguised ... Before the play began the gentlemen ... were impatient and called out often *Fuora, Fuora* [Out Out!],[4] and they made a great noise when they stamp'd and whistl'd.' It is beginning to sound rather rowdy: 'Those that sat in the boxes did frequently spit upon the company in the pit, so that all appeared very rude.' Worse was to come: 'We observed but three acts in the play, which was very immodest and obscene, nothing would please the company, who were ready to hiss, and they [their?] disgust [at] anything that was not filthy.'[5] Little wonder that some English authors expressed grave reservations about continental travel for their young men, who they imagined 'losing their virtue, faith and obedience through exposure to foreign vice'.[6]

Portrait of Francis Willughby painted by Gerard Soest around 1657–60, when Willughby was in his mid-twenties

Portraits of Francis Willughby's parents, Sir Francis Willoughby and Cassandra Ridgeway (the latter by Soest; both undated)

Middleton Hall, Warwickshire: Francis Willughby's birthplace and family home

Green woodpecker (top) and great spotted woodpecker; images obtained by Francis Willughby during his travels (artists unknown)

Black-eared wheatear (top left); lesser grey shrike (top right) and red kite (bottom); images obtained by Francis Willughby during his travels (artists unknown)

Ein Nacht Raab. *A Night Raven Ray. page 279 Tab: 49*

Night heron painted by Johann Walther in Leonard Baldner's unpublished book, purchased by Francis Willughby while in Strasbourg

Paintings of three marine fish (from top: red gurnard, Atlantic stargazer and greater weever) by unknown artists, obtained during Francis Willughby's continental travels

Painting by an unknown artist of a perch in Leonard Baldner's unpublished book, purchased by Francis Willughby while in Strasbourg

A variety of insects by an unknown artist in Leonard Baldner's unpublished book, purchased by Francis Willughby while in Strasbourg

Willughby purchased a tiny parasitical novelty: a flea on a metal chain. Venetian craftsmen advertised their expertise by making and fitting a tiny chain to a flea – notwithstanding its short neck. One was 'a gold chain the length of a finger, together with a bolt and key, [made] with such great skill, and with that accuracy, that they could easily be pulled by the flea as it moved forward. However, the flea, the chain, the bolt and the key together did not exceed the weight of a grain of corn.' Willughby housed his flea in a little box, which he kept warm, allowing the beast to puncture the skin on his hand and feast on his blood each day. John Ray later described the process in the *History of Insects*: 'When they begin to suck they raise themselves almost upright and thrust their snout that comes out of the middle of their forehead into the skin. The itching is not felt immediately, but a little later. When they have filled themselves with blood, they start to eject the bloody faeces through the anus; and they suck in this way for many hours if allowed and throw out their excrement. After the first itch no pain is ever felt.' Willughby kept the flea for several months until it died of cold during the winter.[7]

They were taken to see glass-making by a merchant 'who shewed us great civility', spoke English, and carried them in his gondola to Murano, 'which is some distance from the city, and consists of some islands built with many houses, most of which are inhabited by glass-men'. After observing the furnaces and glass-making process, Skippon was fascinated by the fact that 'The Venetians use glass chamber pots, which are preserved from breaking by being put onto strong stalks.'[8] As well as chamber pots, Murano was also where glass eyes were made, and I wonder whether it was here that Willughby obtained those that now stare eerily out of his seed-cabinet drawers.

On an island 'beyond St Pietro di Castello' (the island of Certosa) they visited a monastery and observed how each of the twenty-five monks had a 'little house and garden by himself' where they kept land-tortoises, 'which lay about seven … eggs apiece in summer time in holes they scrape for them'. Skippon continues: 'These eggs are thus buried in the earth, without any other warmth until next spring, when the young tortoises come forth.' It sounds rather charming, until 'They are counted pretty good meat, and are eaten by these monks.' Once again, Skippon mistook land-tortoises for pond terrapins. Until recently both were widely eaten in certain parts of Italy. The Renaissance chef Bartolomeo Scappi, for example, included recipes for land-tortoise pasticcio and fried tortoise with cucumber in his six-volume cookbook of 1570. An archaeological investigation of a late sixteenth-century monastery near Rome found both tortoises and terrapins on the menu.[9]

The Venetian winter was little better than an English one: 'About the middle of October there was a great storm ... and soon after the winter began, which was very sharp sometimes, and about the beginning of February the weather grew warm again.' Venice continued to enthrall and surprise: 'Some of the Venetian gentlemen are so poor, by reason of their debaucheries and ill husbandry, they go to strangers ... and beg for charity ... there were two that used to come to our lodgings in their gowns and caps, asking our relief with a great deal of humility.' It all sounds suspiciously like some beggars today, for 'some of them do live according to their quality, keeping house, a gondola or two, and yet go up and down begging'.[10]

The diversity of species available in Venice's fish market was breathtaking. I won't repeat Philip Skippon's entire list of sixty species here,[11] but simply quote from his and Willughby's notes:

'Mesoro' *Blennus ocellaris*, butterfly fish, sold in Venice in October, and probably throughout the winter.[12]

'Orada' *Sparus aurata*, gilthead, sold in great abundance.[13]

'Uranoscopus' *Uranoscopus scaber*, stargazer. 'Physicians say that its gall-bladder is good for cataracts', but Willughby could see no reason why this fish's gall-bladder would be more effective than that of any another fish.[14]

'Licetti' *Stromateus fiatola*, butterfish (not to be confused with the gunnel *Pholis gunnellus*, also called the butterfish, but which doesn't occur in the Mediterranean). Willughby commented on its curved lateral line.[15]

'Pesce Petro' *Zeus faber*, John Dory, a species they had previously seen in 1662 in Penzance.[16]

'Sorghe marina' *Gaidropsanus mediterraneus*, Shore Rockling. Ray later saw a similar fish in Chester in 1671.[17]

'Rubellio' *Pagellus erythrinus*, Common Pandora. 'We found the fish much tastier in the winter'.[18]

'Scrofanello' – *Scorpaena* sp., Scorpion fish.[19] Willughby and Ray recognised two species: the small scorpion or Scorpaena and the large scorpion, probably the red scorpion fish *Scorpaena scrofa*, which, as they knew, was capable of inflicting a painful wound, and so-called 'because it pierces and strikes and pours out poison in the manner of a land scorpion'. Later, in their *History of Fishes*, Willughby and Ray recount a story included by Rondelet in his fish book, of a boy who was 'bitten' after innocently placing a scorpion fish inside his shirt. Rondelet's remedy was a mixture of mastic and the fish's liver 'applied to the area around the wound'. In fact, the venom comes not from a bite, but from spines covered with toxic mucous. Commenting on the fish's flavour, Willughby says that the (small) Scorpaena is far inferior to the (red) scorpion, and indeed the latter is the key ingredient in bouillabaisse.

Ray wondered whether the fish called Gattoruggine in Venice is the same as Gesner's Gotorosola fish.[20] He thought that the variant names arose because foreigners were unable to follow the fishermen's pronunciation and suggested that Italians should be consulted on this matter – clear evidence that they weren't always confident about their comprehension of Italian.

In some ways Venice must have been an embarrassment of aquatic riches, and I wonder whether they actually had time to describe and anatomise all sixty fish species? Certainly on any single day the number of specimens they could process was limited, especially if the weather was warm. It appears they had help from one or more servants with the dissections at least. Among Willughby's notes is an account written by Philip Skippon of a dissection of a male stingray accompanied by sketches drawn by one Mr Okely; the dissection ended when Mr Skippon's 'workman' became too tired.

This particular servant may have been a relative of Margaret Oakley, John Ray's future wife, from the Willughby household at Middleton. Although we might turn our noses up at the idea of Willughby not dirtying his hands by dissecting, their servants' role was not really that different from the part played by university technicians or research assistants today. What's more, having someone literally getting their hands dirty, bloody and smelly, while someone else took measurements and notes, seems eminently sensible.

Willughby's strategy with the fish they encountered was almost identical to that adopted with birds: a careful description of the external morphology, with particular attention to the colour, the number and position of the fins, the numbers of rays in the fins

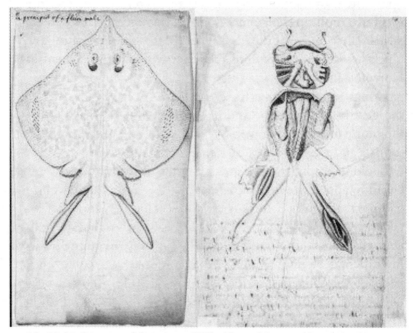

Mr Okely's drawings of the stingray Willughby and Ray saw dissected in Venice.

and so on. Willughby measured the specimens' dimensions, and examined their mouth and teeth; when all that was done, a servant, it seems, opened the fish up so they could see the internal organs: the gut, gonads and liver. Notwithstanding the stingray sketch, we don't know whether they routinely made drawings of the specimens they observed. We take it for granted that on encountering an unknown fish species today – as I did with a curious boxfish on a remote Sri Lankan beach – we save a thousand words by simply taking a few pictures on our iPhone, capturing the way a fish looks and providing the perfect reference from which to make an identification.

Without modern technology, and with insufficient time or talent to create their own high-quality images, Willughby did the next best thing and bought collections of watercolours depicting Venetian fish. This at least gave them something tangible to work from and a wonderful aide-memoire. Those images now lie in the Middleton Collection, many with Willughby's notes scrawled upon them, and several of them very obviously formed the basis of the illustrations in the *History of Fishes*, published after Willughby's death. These images, created by a variety of different artists, include pipefish, seahorses, a writhing conger eel, a bearded rockling, a John Dory, top-knot blenny, garpike, red gurnard and dolphin fish. The paintings are remarkable for their accuracy and beauty – indeed, the next best thing to a colour photograph – but then Italy was the centre of the artistic world at that time, and, no doubt, foreign tourists and travellers created a ready market for such souvenirs.

Looking through this remarkable collection of illustrations, I began to appreciate the magnitude of the task Willughby and Ray had set themselves. I have spent forty years studying birds and I take for granted the fact that I can recognise most species and tell you what family they belong to. The fish are something else, and my experience reminded me of that trick used by those who train people to teach English as a foreign language: they get someone to talk to the class in Norwegian. 'OK?' they say. 'See what it's like?' The sheer diversity of fish body forms, sizes, appendages and colours, as well as their internal structures, meant that classifying fish

– or any other animal group – was far from straightforward. Except for the fact that birds are more visible on a day-to-day basis, I don't suppose Willughby and Ray found them any easier.

}᷆᷉}

It is so frustrating that Francis Willughby's notes no longer exist, for Venice was clearly a wonderful source of specimens and orNitho-logical opportunities. The Venice lagoon was – and still is – excel-lent for birds. In December, for example, some eight kilometres from Burano, they saw 'a multitude of coots and sea-cobs', that is, gulls. They visited a private reed-fringed lake, a *vallè*, with a small island on which lived a keeper who refused them access, 'tho' we requested it very earnestly', for catching fish and fowl. The fish attracted huge numbers of predatory (piscivorous) birds and 'Once or twice a month the owner gives leave to many people, who come in gondola's [sic] and shoot what they can.' The *vallè*, owned by a Venetian nobleman, apparently 'yielded a considerable profit'.[21]

Many of these birds ended up in the Venetian markets, and luckily, Skippon was sufficiently impressed that he left a detailed account, listing twenty-eight species, albeit using only their Latin names: *Arcuata sive Numenius Avis*; *Gallo di Montagna*; *Sardina vel Tardina*; *Alaudae* species, and so on. When my colleague Mark Greengrass worked out what most of these are, the list reveals a rather remarkable range of species: Eurasian curlew, capercaillie, black-eared wheatear, reed bunting, common greenshank, bram-bling, common sandpiper, red-legged partridge, golden plover, smew, spotted redshank, great white egret, merlin, goshawk, avocet, little egret, bar-tailed godwit, red-breasted merganser, red-throated diver, common crane, rock ptarmigan and the 'capo posso' – the lesser red-headed duck, which is probably the ferruginous duck. There aren't twenty-eight species here, because Skippon, tellingly, counted males and females of a few – including the smew and the brambling – as separate species.

It is curious, perhaps, that no such list of birds appears in John Ray's travel journal, but a likely explanation for this exists in the *Ornithology*. That volume contains numerous references to birds

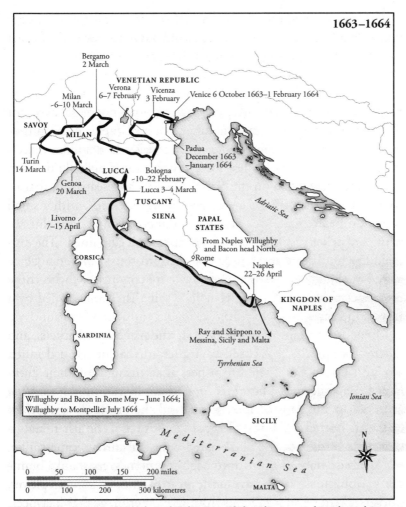

Willughby's continental travels with John Ray, Philip Skippon and Nathaniel Bacon: Italy 1663–4.

seen, described and dissected at Venice, including several that are not on Skippon's list. This suggests to me that Willughby kept the bird notes in Venice and later, with access to these, Ray incorporated the relevant information into the text of the *Ornithology*. Their account of Venetian birds is enhanced by the occasional intriguing observation. For example, in 'the palace of a certain nobleman of the city standing upon the Grand Channel' they get their first, breathtaking glimpse of a gigantic griffon vulture; and bramblings were available in 'great numbers' in poulterers' shops in winter, suggesting that 1663–4 was one in which bramblings made one of their very infrequent mass appearances in this part of Italy. The local bird-catchers called the bird *Pepola* or *Fringuello montanino*, the mountain finch, and these birds were almost certainly caught in the hills not far away. It is also possible that the brambling were taken on the northern side of the Alps where their winter roosts sometimes contain several million birds, attracting all sorts of capture methods – including the use of blow-pipes loaded with clay pellets.[22] Other birds purchased and examined in Venice include whimbrel, water rail, a great skua (dissected by Willughby), tufted duck, goldeneye (that Ray says were common), and perhaps most spectacular of all, a white-tailed eagle that Willughby bought from a fowler.

Obviously, many of these – the waders, herons, ducks and divers – are species that could have been hunted or trapped in the Venice lagoon. But there are others that must have come from elsewhere – the rock ptarmigan, for example, can only have been acquired in the Alps or Dolomites, and the goshawk and capercaillie are both forest birds. Skippon doesn't tell us on what date he saw these birds, but the black-eared wheatear is a summer visitor to northeast Italy so must have been trapped on migration, while the brambling is, as we have seen, an irregular winter visitor from northern Europe. It is obvious from the *Ornithology* that Skippon's list is far from comprehensive and that he simply catalogued species he and his companions were unfamiliar with at home.

The Italians were, and many still are, inordinately fond of eating birds, especially small birds, and the annual north–south migration of millions of birds between Africa and northern Europe provided a rich harvest. Since at least the 1400s, bird-catchers in northern Italy have created large structures referred to as *roccoli* at strategic locations in mountain passes for trapping birds. *Roccoli* usually consist of one or more huge circular hedges up to eight metres high and several tens of metres across. The bird-catchers position their nets inside the hedge, and then use caged decoy birds to pull in the migrants. For some unknown reason, migrating birds find the calls of their own species irresistible and are drawn irrevocably from the sky towards their death. At the centre of the *roccolo* is a wooden or stone tower in which the bird-catcher sits and waits. When sufficient birds have accumulated on the branches of the hedge within the *roccolo*, he hurls from his hiding place a 'bird-racket' (a stick with a flat wicker end) that birds mistake for a hawk, causing them to dive for cover – and into the nets. The hundreds or thousands of *roccoli* scattered across the Italian countryside meant bird-catching – mainly for thrushes and finches – was once practised on an almost unimaginable scale.[23]

Many *roccoli* still exist near Bergamo, maintained either for historical interest or for legitimate bird-catching and ringing (banding), but also – as was apparent when I visited the region – for illegal bird-catching.

It doesn't seem that Willughby and his colleagues knew about *roccoli*, and were happy simply to have access to the spoils. Certainly, many of the vast numbers of dead birds they saw in Italian markets, and especially in Rome and Florence later on, provided an almost unique opportunity to describe and dissect species they hadn't previously encountered. It also allowed them to examine multiple specimens of the same species and see the extent of individual variation in plumage and other features. Together with the fish, they had their work cut out to dissect and describe everything they encountered.

In the spring of 1664 John Ray and Philip Skippon left Francis Willughby and Nathaniel Bacon to travel separately. Willughby and Bacon spent May, June and July 1664 in Rome. Ray and Skippon visited that city later, between 1 September 1664 and 24 January 1665. The bird markets of Rome proved irresistible to both parties, although we have only Ray and Skippon's accounts. Ray was clearly overwhelmed by the sheer numbers of dead birds and the diversity of species on sale: 'partridges of two kinds, the common and red-legged, wood-cocks, snipe, wigeon, teal, bastard plover [lapwing], curlews, quails. Of small birds the greatest plenty I have anywhere seen: as thrushes in winter time an incredible number, black-birds store [many], larks infinite.' And tellingly: 'One would think that in a short time they should destroy all the birds of these kinds in the country.'[24]

Ray also comments on what people were prepared to eat: 'I have seen lying frequently in poulterer's shops, and therefore I presume people eat them, such birds as in England no man touches, viz. kites, buzzards, spar-hawks, kestrels, jays, magpies and wood-peckers', adding that 'Nothing more commonly sold and eaten here, and in all Italy, than coots and stares [starlings].' Later in the *Ornithology*, Ray wrote: 'Stares are not eaten in England by reason of the bitterness of their flesh; the Italians and other outlandish [foreign] people are not so squeamish.'[25]

There is a painting of birds from the 1600s once thought to be the work of the Italian artist Caravaggio, who was active – and dangerous (he murdered someone) – in Rome at that time.[26] The image is a still life of some forty species of exquisitely painted birds laid out as though on a market stall. They include some of those that Willughby and Ray saw in Rome, but there are others too, such as wryneck, kingfisher, golden oriole, barn owl, green woodpecker and smew – all of whose palatability an Englishman might question. Others though are well known to be good eating, including partridges, pigeons, ducks and woodcock. One potentially puzzling feature of the painting is a metal dish on which lie eleven plucked birds, each one – slightly embarrassingly it seems to me – with a single black and white feather protruding from its bottom, where its tail used to be. Naked but not quite, these are northern wheat-ears caught on migration, beneath whose pale skin lies a layer of fat that, had they survived, would have fuelled their migratory marathon. Instead, that fat rendered them a gustatory treat. Old habits die hard, as one of my Italian colleagues told me how as recently as the 1960s, his mother roasted small birds on a spit, allowing the hot fat to drip onto slices of bread that he and his family ate with gusto. It is clear also that the practice of eating small birds in Italy and elsewhere in the Mediterranean continues to this day.

Amid those sad but beautiful painted corpses sits a single live bird: a little owl whose angry yellow eyes stare defiantly at the viewer. Why a *live* owl? The answer is that this was an essential part of the Italian bird-catcher's paraphernalia – as Willughby and Ray saw for themselves. Small birds, especially, would mob the owl – which was tied to a post – drawn either to within shooting range, or becoming ensnared on the horribly adhesive lime sticks that the bird-catchers placed around the owl.

The birds in the painting have been carefully set out; no poulterer's shop ever looked this neat. There's no blood; the birds' plumages are perfect, and instead of lying in a sorry amorphous heap – as I once saw them in a bar in rural Spain – the birds are tidily separated from each other, allowing us to celebrate their beauty and to identify them. In fact, the painting isn't by Caravaggio at all, but just who it was that created this sanitised image of an Italian bird market stall remains a mystery.[27]

Cabinets of curiosities continued to draw Willughby and his colleagues: not knowing what wonders they might encounter, they couldn't resist. Visiting the house of a man known only as 'Rosachio',[28] 'a reputed astrologer, who was a mountebank [charlatan] that sold medicaments in the piazza of St. Marks', they went to see his collection of rarities 'which were kept in pretty good order' with his 'lesser things [small objects] in boxes divided into small partitions, with a wire grate over them' to prevent their loss. Items included the dried tail of a beast that Skippon thought might be a shark, and which Rosachio told them had wings when it was alive. A basilisk perhaps? He also had the head of a 'bachurlars, a bird taken in May about Modena, with much kindness for a man'. The 'bachurlars' or *baciurla* in local dialect, literally means 'simpleton' or 'fool', a term used for several bird species that show 'much kindness for' or no fear of man. In this instance, it may refer to the tawny owl, which in daylight can seem unafraid of people. 'Taken in May' could imply a fledgling owl, which often seems almost indifferent to people.[29]

Rosachio had a small menagerie with a *marmotto* (alpine marmot), which 'sleeps all winter', taken in the Alps; five sorts of parrots in cages, kept warm by a fire; and 'a fine paraquit with a red bill, a very long tail and a black spot and ring about the neck', presumably a ring-necked parakeet.

Another dubious character they encountered was a Flemish-born 'chymist' named Regio, whose trade was secrets (that is, processes), and who said that he had once lived in England with the Duke of Buckingham. For a mere £25 Regio offered to sell Willughby four such secrets: (i) extracting of mercury from lead; (ii) extracting of sulphur from mercury; (iii) fixing sulphur to withstand very high temperatures – although, as Skippon recorded, he 'confessed he wasn't able to fix it completely'; and (iv) making gold volatile. 'Mr Willughby proffered him the much lower price of ten cecchini [zecchini] for these four secrets which he refused to discover [reveal] them for.'[30]

This wasn't the only kind of dubious practice they encountered. The fascination with cabinets and the obsession with finding new specimens meant that travellers were vulnerable to being duped. As early as the mid-1500s, Conrad Gessner warned of 'apothecaries and others who usually dry rays and shape their skeletons into varied and wonderful forms for the ignorant'.[31] Gessner was referring to people such as Leone Tartaglini of Foiano, who in the 1560s and 1570s had a cabinet and a shop in Venice from which he sold all sorts of questionable rarities. These dishonest practices continued and, when Count Lodovico Moscardo produced the catalogue for his Verona museum in 1672, he warned against the 'swindlers and charlatans from Dalmatia' who sold examples of the basilisk in his museum.[32]

Willughby may have been a victim of such a fraudster, for in one of the trays of zoology objects in his seed cabinet I came across the most extraordinary specimen. When I opened that drawer in 2014 for the first time, of all the objects lying there this was the specimen that caught my eye. Despite being a zoologist for almost fifty years I had never seen anything quite like it. The specimen in question is a four-centimetre-long black beetle whose head, thorax and

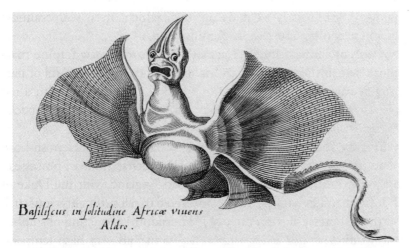

Basilifcus in folitudine Africæ viuens
Aldro.

Basilisk created from a dried ray, illustrated originally by Gessner, copied by Aldrovandi and later, as shown here, by Jan Jonston (1657).

abdomen are covered in hooks and spines – reminiscent of the kind of monster that one would find crawling about in a Hieronymus Bosch painting. The beetle is mounted on a long pin, and looks as though it has just been removed from a display case. I picked it up, and using the pin turned the animal over to look at it properly. It didn't make sense, but was sufficiently convincing that I was unable to write it off as bogus. I showed an image of it to a few entomologists, one of whom realised that its head and thorax were actually the jaws of a fish. He was more specific: they were the pharyngeal jaws of a moray eel. Wow! That was an impressive bit of zoological detective work. I went back to look at the specimen for a second time, and matched up the fish jaws with its backward-pointing curved teeth, but also examined the abdominal spines, some of which were broken, allowing me to see that they were the thorns from a plant – probably hawthorn.

This sophisticated forgery, an exemplar of the mountebank's profession, encapsulates the problem Willughby and Ray faced as they endeavoured to replace fabrications and falsehoods with the truth.

Was Willughby duped or did he buy that counterfeit insect for its sheer ingenuity and novelty value, or as an object lesson – effectively saying: 'Look, this is what we are up against?' It is hard enough trying to identify and classify genuine organisms, without the contribution of animal forgers. The fact that this Piltdown beetle is not mentioned in the *History of Insects* suggests that neither Willughby nor Ray was deceived.

ᚦᚹᚦ

Willughby and his friends left Venice in early December 1664 to spend just over one month in Padua. The university here, founded in the thirteenth century, was one of the great intellectual centres of Europe. They signed up for a series of lectures delivered by Pietro Marchetti, a supremely skilled anatomist and surgeon who had studied under Hieronymus Fabricius ab Aquapendente – better known simply as Fabricius. Fascinated by reproduction, and despite being rather hopelessly loyal to Aristotle's antiquated knowledge, Fabricius conducted pioneering research, including reporting that chickens could store sperm for a full year from one breeding season to another. Fabricius in turn had been taught by Gabriele Fallopio, most famous now for identifying the fallopian tubes (later named in his honour), but his many other discoveries included the third earbone, now known as the 'stirrup' or stapes, and the clitoris. Fabricius revolutionised the teaching of anatomy at Padua by the introduction of a public dissecting theatre – as Willughby and Ray would soon see.[33]

For all his fame, Fabricius, it seems, was a pretty maudlin kind of person, moaning about his salary and the hassle of teaching students – not unlike some of my academic colleagues. His most famous pupil was the English physician William Harvey, who, like many others, spent time in Padua completing his education. Harvey arrived in 1599 and graduated with a doctorate of medicine in April 1602, at the age of twenty-four.

From 1609, Harvey was in charge of St Bartholomew's Hospital in London, and from 1618 was 'physician extraordinary' to James I (and later Charles I). Like his Paduan predecessors, Fabricius and Fallopio, Harvey was interested in 'generation' – reproduction and embryo development. The challenge he set himself was to understand the crucial process of what we now refer to as fertilisation, but in his case, the role of semen in the generation of new life. Recognising how difficult that would be using quadrupeds or humans in this research, he focused his attention on birds whose external embryo development arrangement – otherwise known as the egg – was, he thought, more tractable.

Harvey conducted experiments with broody hens that he kept in his house, disproving Fabricius's notion that they could store sperm and produce fertile eggs for as long as a year after separation from a cockerel. Hens, like all female birds, can store sperm, but as Harvey demonstrated, only for a maximum of thirty days. Only? It is still pretty remarkable and utterly different from what happens in humans and indeed most mammals. Harvey was super-smart and when he discovered errors in his supervisor Fabricius's work, he said so. In many ways, he was the archetypal academic: an astute, rigorous and clear thinker; he knew the literature and understood what needed to be done to solve the mystery of fertilisation. What he didn't know was that he would need a microscope to do so. The hand-lens he relied on simply wasn't sufficient to allow him to see the minute male and female gametes, and he failed to make the breakthrough he was hoping for. Frustrated and disappointed, his notes on generation lay unpublished – and unknown – for many years until his friend George Ent took them and arranged for their publication in 1651. That book, *On the Generation of Animals*, was to serve as an invaluable source of information about reproduction when Willughby and Ray started writing their *Ornithology*.

Harvey wasn't interested only in bird and mammal reproduction, for he also studied insect generation, but his notes were lost when in 1642 his Whitehall lodgings were ransacked by Parliamentary soldiers during the Civil War. Had his notes survived and been included – as they surely would have been – in his *Generation* book, they would have influenced and inspired Francis Willughby and helped John Ray when he finally started to pull together all of Willughby's entomological results.

Pietro Marchetti, whose lectures Willughby and Ray attended, was Italy's most celebrated surgeon. His expertise spanned neurosurgery, anal fistulas and gunshot wounds. He was probably kept busy with the latter at least, for Padua was a dangerous place as Ray noted: 'The citizens and strangers here dare not stir abroad in the dark, for fear of the scholars and others, who walk up and down the streets most part of the night, armed with pistols and carbines.'[34]

Ray and Willughby attended a dissection of a woman under Marchetti's instruction in the surgeon's home. This came about because at the time of their visit Francis's relative, Charles Willoughby, was studying in Padua to become a physician. It was he who told Ray and Willughby about this particular anatomical event and arranged for them to attend. There are few details of what must have been an edifying experience, other than the fact that it occurred over ten days in mid-December, when the cooler temperatures helped to keep cadavers 'fresh'. Even more enlightening, on Christmas Day 1663, Willoughby and Ray were invited to observe a Caesarean operation performed by Marchetti's son, Antonio. When I first read of this I hoped for the sake of both the mother and the spectators that the procedure was successful. But Ray's notes explain that the mother was already dead, and that the extracted infant lived only a few days. It was once thought that Julius Caesar had been born this way, hence the name of the procedure, but it appears not to be true. However, full-term foetuses whose mother had died were sometimes saved by being cut from the womb – although not in the case witnessed by Willughby and Ray. Only in the nineteenth century were Caesarean sections (as they became known) used routinely to save the lives of both mothers and infants.

More dissections followed after Christmas, including a caper-caillie and a female hare whose stomach contents Ray at first thought smelled of honey, but on reflection he realised it might have occurred because the animal had consumed material from a fir tree. The details of these anatomical adventures don't appear in Ray's published account of their travels, possibly because they might have been considered inappropriate. Instead, the notes were found among Ray's papers after his death by Samuel Dale, who in 1706 published them in the Royal Society's *Transactions*.[35]

It was while in Padua that Willughby purchased a collection of dried plants – a herbarium – from the famous physic garden there. The plants were pressed and dried, attached to sheets of paper, labelled and bound between covers like a book. Just as with the bird and fish paintings that Willughby already owned, the herbarium provided an invaluable source of reference material. In truth, however, the herbarium was more equivalent to a physical collection of insects and the dried skins of fish, reptiles and birds, with the added advantage that the botanical specimens survived the ravages of time rather better than zoological material. When Willughby and his friends visited Padua's wonderfully geometric physic garden – which still exists – it was already over a century old with a splendid reputation for the roughly 2,000 plant species it contained. It was probably because of the garden's reputation that both resident and visiting physicians were keen to possess their own Paduan herbarium – a desire encouraged no doubt by the gardeners happy to supplement their

The botanical garden at Padua was the first in the world, and this is what it would have looked like when Willughby and Ray visited it.

income. Willughby had seen a Paduan herbarium previously when he and his colleagues were in Altdorf and he presumably realised that acquiring one for himself was a possibility. However, the herbarium that Willughby purchased in Padua, and which is now part of the Middleton Collection, is puzzlingly limited in both the number of specimens – just sixty-nine – and their taxonomic diversity: they are mainly plants of the daisy and cabbage families.[36]

In early January they returned to Venice for a further twenty-six nights, after which on 1 February 1664 the four companions finally said farewell to that city and travelled westwards towards Vincenza. After arriving there on 3 February and taking note of its extensive silkworm industry and its 'rich and gustful' wines, they travelled the ten kilometres to the famous cave of Costozza where they visited an enormous roost of hibernating bats. In the same cave 'in some standing waters they also encountered a curious animal' that John Ray describes as 'a kind of fish or rather insect [that is, invertebrate] which they call Squillae Venetianae, i.e. Venice shrimps, but they are that sort which naturalists call *Pulices marini* or *aquatici*, i.e. sea-fleas or water-fleas'. Ray is referring here to an eyeless, two to three centimetre-long shrimp-like creature with the name of *Niphargus costozzae*.[37] In Ray's day the local name *squilla* meant 'shrimp', creating the potential for confusion with the mantis shrimp *Squilla mantis*, common on the muddy bottom of the Venetian lagoon and considered a delicacy, but strictly marine. There's a fine dual – dorsal and ventral – portrait of *Squilla mantis* among the Venetian 'fish' images purchased by Willughby, but no image of *Niphargus costozzae*.

Making their way on to Verona they visited Lake Garda, which Ray tells us 'furnishes the city with plenty of excellent fish' including an unusual species 'of the trout kind, called *Carpione*, peculiar to this lake. Those we saw were not a foot long, of the fashion of a trout.'[38] This species, known as *Salmo carpio*, is related to the salmon and is rather like the charr that Willughby had seen previously in the Lake District and in Wales, and endemic to certain lakes. The *carpione* is unique to Lake Garda, but only just since its numbers have declined catastrophically since the time of Willughby's visit. Wondering perhaps whether Lake Como was home to other endemic fish, Francis left his three colleagues near Milan and went north alone to see for himself. When they all met up again in Milan, Willughby told them that the lake at Como 'affords a great store of fish, viz. 1. Bottatrice; 2. Agone, which are catch'd best in the darkest nights; 3. Pisce piso which hath a thorn or prickle on every scale',[39] but presumably, no *carpione*.

Because fish like the *carpione* and charr have evolved independently in different lakes, they were, and continue to be, a nightmare to classify: are those from each lake a separate species, or should they all be lumped together? The debate continues. Of the three Lake Como fish mentioned by Willughby: the *agone* is the twaite shad; the *bottarice* is the burbot *Lota lota*, an unusual bottom-living species and the only freshwater cousin of the cod, and now extinct in Britain. Its name refers to the single sensory barble that dangles beneath its chin. Like its marine relative, the burbot is excellent eating. The *Pisce piso* is probably what is known now as the pigo *Rutilus pigus*, a roach-like fish in Lake Como that in the breeding season develops a covering of impressive thorns or spines on its scales, and unlike Willughby's beetle, is a genuine beast.

As an undergraduate Willughby had written about Ulisse Aldrovandi's museum in his commonplace book and no doubt fantasised about seeing it for himself. Considered the father of natural history, Aldrovandi studied mathematics and medicine at Bologna, where he later created the botanic garden. In 1549 he was locked up for heresy – for denying the Trinity – and kept under house arrest until the following year. It was during his incarceration that he turned to natural history. The eventual result was a vast museum of curiosities and a set of impressively large volumes describing much of the world's *biota*, as it was then known.

So it was that in mid-February 1664 Francis Willughby and his three friends were taken to the Palazzo Pubblico in Bologna and

shown around Aldrovandi's collection by the curator, Professor
Ovidius Montalbanus.[40] With some 7,000 plants and 18,000 other
specimens in six rooms, they must have been overwhelmed. Philip
Skippon noted some of what was on display, including: malformed
fruit (how did that occur?); a puppy without a head that died soon
after birth (really?); a 'notable signature of a spider's web'; a dragon
or snake with wings and legs; the antlers of an 'old stag that had
done branching, and began to degenerate into rough extuberances';
a picture of a hairy girl born of two hairy parents; a hen's egg shaped
like a gourd, and so on. It wasn't entirely Aldrovandi's handiwork,
since, after his death, the collection had been purchased in the 1650s
by the rather sad-faced nobleman Fernandino Cospi, who merged it
with his own and then presented it all to the state in 1660.

One of the rooms, presided over by a lively portrait of Aldrovandi
himself, contained his books: ten folios of plants and seven of birds,
fish and insects painted in vibrant watercolours. Willughby's eyes
must have lit up at the sight of these images, most of which still
survive.[41] Because Aldrovandi had been able to employ the very
best artists in Italy, the paintings are both beautiful and accurate. It
was one thing to see the monochrome engravings in Aldrovandi's
published works, as Willughby and Ray had undoubtedly done,
but quite another to see the vividly coloured originals on which
those black-and-white images were based. The fish are painted better than the birds, but then fish don't differ as much in their posture
between life and death, although the colours of some species fade
rapidly after death. Regardless, this must have been the standard to
which Willughby and Ray aspired as they thought ahead to the way
their own publications would be illustrated.

As great as Aldrovandi was, he was also part of the problem
and one of the main – but not sole – reasons that Willughby
and Ray decided natural history needed a makeover. As is clear
even from Skippon's short list of specimens, Aldrovandi was an
indiscriminate and undiscerning collector. Anything odd from
the natural world would do, and Aldrovandi's was an age of exu-
berant knowledge: the more connections he could find to a specimen the better. It makes – assuming you can read Latin – for

extraordinarily tedious reading. Undoubtedly, in the late 1500s and the early 1600s such voluminous, discursive and wide-ranging accounts masqueraded as great scholarship, and while Aldrovandi must have been absolutely driven to achieve what he did, his was an unenlightening, shotgun approach to understanding the natural world.

Subsequently, a major task for Willughby and Ray was to match the animals and plants included in Aldrovandi's books with what they had seen for themselves. It wasn't always easy and it would have been much more straightforward had they had copies of those colour illustrations to refer to. I know what it must have felt like for I have struggled to identify some of the curiously named animals and plants listed in Skippon and Ray's journals. In many cases it was impossible for Willughby and Ray to unequivocally establish what species of bird or fish Aldrovandi was writing about; sometimes he made mistakes – in part because he was more focused on quantity than quality, and like many other naturalists, often assumed the two sexes, or adult and immatures, to be different species.

Even so, Willughby's visit to Aldrovandi's museum must have been among the highlights of the entire continental journey.

From Bologna the four travelled to Milan where they came across a censored version of Gessner's *History of Animals*, one of the titles banned by Pope Paul IV in his *Index Librorum Prohibitum* of 1559. The Index didn't explain why certain books were banned, but in his *Bibliotheca Universalis* – the first list of all books since printing began, published between 1545 and 1549 – Gessner had publicised, and by doing so had promoted, Protestant works. This did not endear him to the Catholic Church and his *Bibliotheca Universalis* also appeared on the list of books banned by the Pope. As Skippon reported, on the title page of Gessner's *History of Animals* someone had written 'Damnati authoris [damned author] &c.', and he tells us that 'all those notes which Gessner calls superstitious and magical were blotted out'.[42] As the historian Mark Greengrass says, this kind of censorship 'may have confirmed Willughby's worst suspicions about the relationship between natural philosophy and the Counter-Reformation Roman Catholic Church'.[43] Curiosity about the natural world was dangerous for religion, and examples of such papal 'censorship' can be seen on the Internet.[44] Part of the motivation for censorship in this case may have been due to the strained relationship between the fervently Catholic Aldrovandi, who quoted extensively from the strictly Protestant Gessner – but without mentioning him by name. Aldrovandi anonymised Gessner, referring to him as 'ornithologus' (bird man), presumably because he feared Catholic sanctions. Another explanation is that Aldrovandi felt threatened by Gessner, who wrote to him continuously questioning his natural history writing: Aldrovandi never answered Gessner's letters.[45]

From Milan, Willughby's party travelled to Turin, Genoa, Lucca, Pisa and to Livorno where, on finding a Dutch boat ready to sail for Naples, 'we put ourselves aboard her'. With the wind being against them, the sea passage took five long days. Luckily, their arrival in Naples coincided with a weekly meeting of the 'virtuosi or philosophic academy … in the palace of that most civil and obliging, noble and virtuous person, the marquess D'Arena'. This was Andrea Concublet, who had reopened the *Academy of Investiganti* after it had been closed as a result of the plague (which killed over two-thirds of Naples's population) in 1656. Willughby and his colleagues admired Concublet's natural history museum, and, together with an audience of sixty or more, observed a practical demonstration of 'water ascending above its level in slender tubes' followed by discussion and the reciting of discourses composed about particular subjects that had been allocated to the participants the week before. Willughby and his party were deeply impressed and commented: 'A man could scarcely hope to find such a knot of ingenious persons and of that latitude and freedom of judgement in so remote a part of Europe.' As this suggests, Naples was a city of great cultural vitality and, being in touch with the Royal Society, was one of the main academic centres of Europe. Willughby and his friends found the local virtuosi to be up to date with the works of other European scholars and were 'very much pleased and satisfied with the conversation and discourse'.[46]

}⋏{

At Naples the party split up, with John Ray and Philip Skippon continuing south to Sicily and Malta, and eventually back to Venice, while Francis Willughby and Nathaniel Bacon returned north to Rome. Because Ray and Skippon, whose journals still exist, travelled together, we have a detailed account of their journey, but – frustratingly – almost nothing for Willughby and Bacon whose accounts are now lost.

Ray and Skippon botanised as they travelled southwards, finding many new plants, but also, on returning to Rome in September 1664, a new species of bird: the citril finch. Carefully distinguishing it from the similar siskin and serin, Ray tells us how it was known as the *verzellino* in Rome, and was 'nurtured in cages for the sake of its singing'. Also in Rome, Ray and Skippon were able to visit the so-called paper museum of the wealthy and politically astute nobleman Cassiano dal Pozzo. His museum was an ambitious attempt to represent the entire natural (and unnatural) world visually, through an enormous and magnificent collection of superb paintings created by the most accomplished artists of the day. The English physician, George Ent, met dal Pozzo during a visit to Rome in 1636 and afterwards corresponded with him. Dal Pozzo later sent Ent examples of petrified wood that were exhibited at the Royal Society in London and helped to fuel the debate about the origin of fossils.

One of dal Pozzo's other claims to fame was writing and producing one of Italy's great seventeenth-century bird books, albeit under someone else's name. At this stage he could only dream of being a member of Italy's scientific elite, the Accademia dei Lincei (Academy of Lynxes), but realised that were he to present them with a definitive work on ornithology, it might grease the wheels. Many years previously, in 1601, Antonio Valli da Todi had published a small, nicely illustrated volume on birds – mainly species that could be kept in cages for their song or appearance. Rather deviously, dal Pozzo recognised that with a few additions Valli da Todi's work could be updated and passed off as a new book. He persuaded Giovanni Pietro Olina to front it. They pulled one of the original artists, Antonio Tempesta, out of retirement and persuaded him to touch up some of his original images, and they enlisted another brilliant bird artist, Vicenzo Leonardi, to create some new paintings. Dal Pozzo then presented the volume – the ultimate guide to the increasingly popular practice of bird-keeping – to the Academy in 1622, with these words: 'I send you this bird book produced by one of my affiliates as a tribute of my respect, as evidence that the evidence I gather with limited effort and money can contribute towards this field [of science].' The ploy paid off: the combination of the bird book and his remarkable collection of natural history images got him elected to the Accademia dei Lincei.

Olina's book *Uccelliera* (*The Aviary*) was a great success. The images were superb, as Willughby and Ray recognised, and they later made copies of them for their own book. Along with those from Thomas Browne, they are among the best in the *Ornithology*.

Olina's book also provided some intriguing bird biology. In the English edition of the *Ornithology*, Ray included this paragraph: 'It is proper to this bird [the nightingale] at his first coming [arrival from migration in spring] (saith [says] Olina) to occupy or seize upon one place as its freehold, into which it will not admit any other nightingale but its mate.' This is the first truly explicit statement that birds defend a breeding territory and it also suggests that it was Giovanni Olina who discovered this fact. But it was actually Valli da Todi who said it first, and it was he who – in his own book that dal Pozzo and Olina had so blatantly plagiarised – also pointed out that nightingales accomplish this defence by singing. Territory or 'freehold' is one reason why birds sing. Together, these extraordinarily significant discoveries in bird biology were ignored for another three centuries. Valli da Todi's ornithological insight was based on the fact that he was a bird-catcher with a great deal of first-hand experience of nightingale behaviour (essential, of course, if he was to stay in business). Ray recognised that there was something significant in Olina's stolen phrase about a 'freehold', but, with limited experience of actually watching birds himself, he was

Nightingale (top left), brambling (top right), skylark (bottom left) and francolino (bottom right) from Olina's (1622) Uccelliera.

unable to make the link with song, nor recognise the general significance of territoriality.

In Rome, Ray and Skippon met up with Sir Thomas Browne's son Edward, who was on his own continental tour. Edward wrote to tell his father of the encounter, but also to explain how Ray's collections of 'plants, fishes, fowls, stones and other rarities' from Germany and northern Italy had been sent from Sicily back to Britain accompanied by Skippon's servant, only to be intercepted and stolen by Barbary pirates; the unfortunate servant – like many white Christians intercepted at sea – was now a slave in Tunis.[47]

The details are vague, but Willughby and Bacon seem to have left Rome in the summer of 1664 and travelled back to northern Italy where Bacon had intended to meet up again with John Ray. But Bacon contracted smallpox and was able to join Ray and Skippon only in March 1665 when they were in Venice. Bacon then returned to England from Genoa on 20 April of that year. Willughby meanwhile must have seperated from Bacon soon after he was taken ill, and made his way westwards into southern France.

Ray and Skippon continued north from Rome, arriving in France in July 1664, and via Lyon, Grenoble, Orange and Avignon, reaching Montpellier later that summer. Like Padua in Italy, Montpellier was an intellectual centre, with an outstanding reputation for the study of botany. It was here that Ray, and Willughby who arrived later, met several other English virtuosi, including Sir Thomas Crew (who proved to be a source of some excellent bird illustrations), Nicolaus Stensen (known as Steno, the great Danish anatomist), their friend Francis Jessop from Broomhall in Sheffield, and perhaps most significantly of all, Martin Lister, with whom Ray became good friends.[48]

The Protestant expatriate community of Montpellier found themselves in a slightly unusual situation. Forbidden by university regulations from attending lectures or matriculating in medicine, they were expected to pay for private tuition if they required specific teaching. The ex-pats therefore rather had to fend for themselves, but that seems to have been more than acceptable.[49]

It was from Crew in Montpellier that Willughby and Ray acquired a rather beautiful image of a grouse-like bird known only as 'Le Jangle de Languedoc'. Its identity was something of a mystery and at the top of the painting Ray wrote: 'A bird of passage: this bird is I suppose the same that is figured and described by Olina under ye title of Francolino though ye colours differ being corrupted by the paintor [sic] to make the bird show beautiful.' In other words, Ray thought the bird might be an over-coloured rendering of

Olina's *francolino*: the hazel grouse. Its name 'angel of Languedoc', however, provides a clue, as does the comment, presumably from someone local, that it was a bird of passage or migrant. The image is in fact that of a pin-tailed sandgrouse, a species that occurs very locally during the summer months in the stony desert-like area of Le Crau some sixty kilometres east of Montpellier. Indeed, this is the bird's most easterly outpost on the continent and explains why Willughby's party never encountered it in the Italian bird markets. As the painting shows, despite its bright colours, the mottled plumage is highly cryptic, making the sandgrouse difficult to see when crouched on a stony substrate. In the air sandgrouse fly fast and high – hence 'angel' perhaps – posing an identification challenge even for today's birdwatchers. Little wonder that neither Ray nor Willughby had seen one alive, let alone a dead one, and they continued to be puzzled by the bird's identity. So uncertain were they that the pin-tailed sandgrouse is absent from the *Ornithology*.

Gessner had previously referred to a migratory bird 'near Montpellier ... commonly called an angel' whose image he had been sent by Guillaume Rondelet. But Gessner made the mistake of assuming the bird to be the same as one known to the Arabs as the 'alchata', and he decided it was a pigeon. Pierre Belon, the French naturalist, doesn't mention the species at all, and Aldrovandi simply plagiarised Gessner, so no progress there. Later, in 1668, Walter Charleton, who knew both Willughby and Ray, published some images of birds, including the 'alchata'. This bird it turns out was copied from Thomas Crew's painting of the sandgrouse, and, unlike Ray, Charleton picked up on the connection between alchata and angel, which are onomatopoeic, reflecting what today's field guide notes is the bird's nasal, hard-grating *rreh-a* voice.

The sandgrouse puzzle persisted for a further century until the Comte de Buffon connected all the pieces of the puzzle to make the correct identification.[50]

⸙

On 1 February 1666 an Anglo-French shipping dispute in the English Channel resulted in Louis XIV giving any Englishman in France three months to leave. This brought our naturalists' genteel wining and dining in Montpellier to an abrupt end, with Ray returning to England by April of that year.

Earlier, after separating from Nathaniel Bacon in northern Italy in August 1664, Willughby had travelled via Montpellier into Spain accompanied only by a servant. Willughby's account of this lonely journey was subsequently paraphrased and published by Ray as a kind of appendix to his own account of the entire continental expedition.[51]

Some eighty kilometres west of Montpellier, at Narbonne and unable to ride because of a sore leg, Willughby decided to 'go forward by sea'. But in the week they waited for a fair wind, his leg healed thanks to a plaster of diapalma (a concoction of oil, hog fat and litharge of gold – gold and red lead). Purchasing two mules for 'five pistoles apiece' Willughby and his servant headed off towards Perpignan, crossing the border into Spain near Baynuls-sur-Mer on 31 August 1664, 'without danger, searching or any trouble at all'.

Willughby must, I suspect, have carried with him Martin Zeiler's compact Spanish itinerary with its folded map of the Iberian Peninsula. Zeiler pulled no punches saying that there was little worth seeing in Spain other than the royal palace at Madrid and San Lorenzo de El Escorial, some forty-five kilometres to the north-west of that city. Willughby knew Spain would be tough, for another travel guide he had consulted before setting out from England was explicit: prospective travellers would require a 'good store of Phlegme and patience'.[52] It was all true, and Willughby soon began to hate Spain.

He started off as he had when with his companions, recording various things of interest; the coral fishing at Capo de Creux (Cruess), an aqueduct made by the Moors, an amethyst mine, and an account of the manufacture of sugar from cane. But over successive days you can feel his enthusiasm draining away. In the market at Valencia, the locals, unused to strangers, shouted and threw fruit peel at him. At the university there, Willughby attended some lectures but, discovering that the new science was unknown, concluded that Spanish academics were a hundred years out of date. There were a few highlights, including the discovery of chocolate in

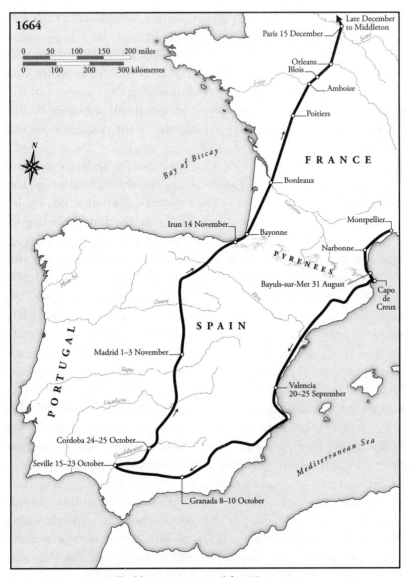

Willughby's journey to and from Spain, 1664.

Seville, Signor La Stannosa's 'famous museum' at Xuesca (Huesca) that Willughby was told would require several days to see completely, and 'a palace of the moors [at Granada – the Alhambra] that well deserves a journey of a dozen leagues', and in Madrid he saw the Palacio Real, El Pardo and El Escorial.[53]

On 1 November, as Willughby travelled north across the Sierra Morena near Toledo, he left behind the Mediterranean warmth and experienced a 'hard frost and pretty thick ice'. Travelling was tiring and tedious and he and his servant were forced to travel at night to avoid paying bribes or being searched by the customs agents that seemed to be everywhere. His natural history observations were few, noting only 'a very good breed of falcon' – presumably the peregrine, or just possibly Eleonora's falcon that might then have bred on the Spanish mainland – at Alicante. In the Valencia markets he found an abundance of swallows and sand martins, which the locals referred to as *papilion di montagne* or mountain butterflies, a lovely name for the sand martin that sits uncomfortably with the idea that they were there to be eaten. It was also in a Valencia market that Willughby unwittingly made an important discovery. Looking over the piles of dead birds for sale, he found and purchased a small and unfamiliar rail. His meticulous method of description that so annoyed John Ray now allows us to identify it as a spotted crake, which may have either been part of the small breeding population in Spain or more likely a bird on migration. However, as Ray was later preparing the *Ornithology*, he decided that Willughby's distinctive Valencia crake must, after all, be the same as the Baillon's crakes they had described previously in Italy: 'Yet I perswade my self that both these descriptions [that is, Ray's of Baillon's crake and Willughby's of the spotted crake] are of one and the same species of bird, differing either in age or sex.'[54] The crake is, I suspect, the first ever description of the spotted crake, a fact that seems to have been overlooked and another feather in Willughby's cap.

Willughby's list of what was wrong with Spain is a long one. The country, he writes, is:

in many places … very thin of people, and almost desolate. The causes are: (i) a bad religion; (ii) the tyrannical inquisition; (iii) the multitude of whores; (iv) the barrenness of the soil; and (v) the wretched laziness of the people … very like the Welsh and Irish … For fornication and impurity they are the worst of all nations, at least in Europe; almost all the inns … having whores who dress the meat, and do all the business. They are to be hir'd at a very cheap rate. It were a shame to mention their impudence, lewdness and immodest behaviours and practices.

The contrast with what had gone before in France and Italy could hardly have been greater. Willughby was weary, lonely and unstimulated. Reaching Seville on 16 October 1664, he tried hard to secure a passport for Portugal, which he hoped to reach via a trip to Tangiers and from there by boat to Lisbon – the best way into Portugal he said – but a combination of factors, including his 'mule's ill fortune' (presumably lameness) scuppered this particular plan.

Two letters awaited his arrival in Seville. One was from John Wilkins, who wrote to persuade him to travel – at the Royal Society's expense – to Tenerife to make astronomical observations of the transit of Mercury across the sun, from the 12,000-foot summit of Pico del Tiede. The second letter informed Willughby that his father was seriously ill.

Had Willughby made the trip to Tenerife he might have discovered some new birds, including the distinctive and beautiful blue chaffinch and two species of pigeon (Bolle's and Laurel pigeon), all of which occur only on that island.

Abandoning the idea of visiting Tenerife, Willughby hastened – albeit on his despised mules – for home via Toledo, Madrid and Burgos, arriving on Spain's northern shore at Irun on 14 November 1664. It is here that Willughby's account, published by John Ray, ends – almost. A letter from Willughby to Ray places Willughby in Paris by 15 December. Briefly describing his route from the Spanish border – Bayonne, Bordeaux, Poictiers (Poitiers), Amboise, Blois and Orleans – Willughby writes with no punctuation, and presumably in an even greater rush than usual:

> [I] got hither [Paris] a fortnight since this journey of almost a thousand miles [1,600 kilometres] I came all alone having agreed with my merchant [presumably someone he had met en route] ... and I thank God [e]scaped very well along but at Vittoria [Vitoria-Gasteiz] and San Sebastian ... was basely troubled with searchers [security men, presumably corrupt].[55]

It must have been with a sense of both relief and achievement – after a year and a half away – that Francis arrived back at Middleton just before Christmas 1664 to find his father still alive.

7

Back at Middleton

Willughby's father, Sir Francis Willoughby, it seems, was a man of sweet temperament and great virtue, and it was he who had saved the family from financial ruin in the early 1600s. As his granddaughter Cassandra later said: 'He, by his seasonable vigour and understanding, retrieved, repared [sic] and restored the not only shattered, but almost ruined estate of the family.'[1] It was he, of course, who paid for his son's education and travels and accepted Francis's decision to lead a life of study.

However, when Sir Francis died the following year, in December 1665,[2] Francis had no option but to take over the running of the estate, which he did alongside his studies. He had a strict routine, working at his research in the morning between getting up and eleven o'clock when his mother oversaw family prayers. After this he walked: 'which for the most part was all the time he spared from his studies in the whol [sic] day, for as soon as he had dined he would again return to his books', that is, 'unless company came to viset [sic] him who he thought himself obliged to entertain'.[3] I was struck by the similarity with Darwin's rigorous, somewhat antisocial daily routine in which work was broken by a restorative, contemplative walk, a regime only possible among academic men of independent means.

While in London, during May 1665, risking the plague that was beginning to become prevalent once again, Francis had received a letter from his father urging him to 'match himself', that is, to find a wife. After his father died, Francis's brother-in-law, Sir Thomas Wendy, continued to encourage him, saying: 'now that he was become the head of the house he hoped that he would lose no time, but settle himself and bear a share in publick imployment'.[4]

Looking after the estate was one thing, marrying was another. As his daughter later wrote, Willughby deferred marriage 'lest his being engaged to a family should prevent his pursuing those studies which he was then so very intent upon'.[5]

His studies at this time included constructing classifications of both birds and insects, an important impetus for which was this request from his colleague John Wilkins: 'I must desire your best Assistance for the regular Enumeration and defining of the Families of Plants and Animals ... I [do not] know any Person in this Nation who is so well able to assist in such Matters as your self, especially if we could procure Mr Ray's Company to join in it.'[6]

Francis and Wilkins had first met in Oxford in 1660 when Wilkins was Master of Wadham College and Willughby was reading in the Bodleian Library. Then, during his brief tenure as Master of Trinity, Wilkins had also met John Ray and knew that between them, Willughby and Ray possessed the knowledge and skill to help him pull off his ambitious plan to create a universal language of science. The three of them had discussed it when they first met, and, indeed, it was Wilkins's grand scheme that set Willughby and Ray on their course of classifying and 'ordering' the natural world. It was also Wilkins who encouraged them to collect and classify words in different languages during their travels.

By 1666 Wilkins's plan was beginning to take shape and he needed some expert scientific input on the business of classification. His idea was to use tables demonstrating the natural order and relationships between things to create a universal language of science, facilitated by the use of 'real characters' or symbols similar in principle to the notations used in music or mathematics.

The ambiguity and confusion over the names of things, whether they were fowl, fish or flowers, was very real. Since a primary aim of the new science was to encourage clarity by avoiding ambiguity, it isn't surprising that Willughby and Ray were happy – initially, at least – to support Wilkins's project.

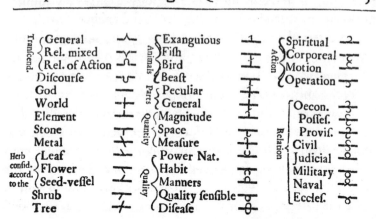

Chap. I.   *Concerning a Real Character.*   387

Some of Wilkins's 'real characters' from his Essay towards a Real Character *(1668).*

Soon after receiving Wilkins's request in October 1666, Willughby started in earnest to prepare his classificatory tables of insects, fish and birds. The insect classification tables that he subsequently sent to Wilkins were constructed from work already completed before the continental journey, for Willughby and Ray began making observations on insects while still at Cambridge; and by 1661, Willughby had a sufficient grasp of the topic to send Wilkins a remarkable, albeit brief, overview of insect biology. The account was never published and the original lost, but a copy in Henry Oldenburg's hand survives among the Royal Society's Classified Papers. It was read at the Royal Society on 20 November 1661 and provides a general account of insect life cycles, including development and metamorphosis; egg design; egg-laying behaviour; and the puzzling fact that some insect larvae often gave rise to other organisms. As an overview, it seems very basic by today's standards, but in an area where knowledge was so limited, Willughby's insect adventures broke new ground. His findings were exciting and within a decade others were eagerly following his entomological lead.

Willughby and Ray's fascination with insects had started in August 1658 when they reared ten large white butterfly caterpillars in a box. After feasting for several days on the turnip and cabbage leaves provided, the caterpillars attached themselves to the sides of the box in preparation for pupation. Soon after, seven of the caterpillars 'revealed themselves to be viviparous', literally, giving birth to live young, or 'vermiparous', giving birth to worms, 'for out of their backs and sides there burst out very many little creatures of the class of maggots'. Between thirty and sixty maggots emerged from each caterpillar, after which the caterpillar expired. Immediately after their birth, the maggots 'were opening a weavers shop' – meaning that they were weaving for themselves silk caskets or dwellings alongside the caterpillar's corpse. Some six weeks later, in October, a little fly emerged from each silken cocoon. Of the caterpillars that did not produce maggots, their pupae created cabbage white butterflies the following April.[7]

What on earth was going on? Why should a species sometimes give rise to its own kind, but at other times produce a completely different type of animal? Willughby and Ray were mystified, but intrigued. They were also careful not to jump to any conclusions. Instead, they posed the testable hypothesis that the large healthy caterpillars gave rise to butterflies, but the weaker and thinner ones, for which 'nature was unable to achieve her original intention', had to resort to producing 'a less perfect little creature, namely a fly'.[8]

In the mid-1600s it was widely believed that one species could transform itself into another, and that life was sometimes created spontaneously, with maggots emerging from corpses, fleas from dust and flies from filth or fermenting fruit. We have Aristotle and his followers, such as Pliny the Elder and Marcus Terentius Varro, to thank for these two specious notions, both of which proved difficult to dislodge. Aristotle, for example, suggested that redstarts turn into robins in the winter. This is not quite such a silly idea as it seems when you know the circumstances. Redstarts were (and still are) summer visitors; robins, however, were generally only winter visitors to coastal Greece where Aristotle once lived. As redstarts disappeared, robins seemed to replace them. Aristotle's deduction about the cause of this change in the local avifauna was shown to be wrong by William Turner in 1544 when he reassessed Aristotle and Pliny's (largely plagiarised) ornithological observations. This transformation could not be true, he said, because in Cambridge he had seen both species side by side.

Turner was a forerunner of the new science and one of Willughby and Ray's most important ornithological predecessors. It wasn't only animals that were thought to transform themselves; as Ray points out in his catalogue of Cambridge plants, it was once believed that turnips could grow from cabbage seed. He then adds: 'In fact we have learned that a fairly close relationship exists between these two plants from the viviparous caterpillars that arise on cabbages, which feed off the leaves of turnip no less greedily than off those of their own cabbage, although they disliked most other leaves that we have given them as food.'[9]

Aristotle promoted the idea that animals, especially invertebrates, could be spontaneously spawned from dirt. Oysters emerging from the seabed were his prime example and it was also a convenient explanation for European eels – whose reproduction remained a mystery until the late 1800s. Spontaneous generation was finally laid to rest by some clever experiments conducted by Louis Pasteur in the nineteenth century.[10]

John Ray, however, was among the first to nail his colours to the mast and declare spontaneous generation implausible: 'All insects are the natural issue of parents of the same species with themselves.'[11] He wrote to tell the Royal Society this some thirteen years after his and Willughby's initial observations, and it seems impossible that they did not discuss spontaneous generation as they watched those maggots burst out of the cabbage white caterpillars. In his letter Ray acknowledged that it was the ingenious experiments by the Italian naturalist Francesco Redi, published in 1668, that finally convinced him. Tantalisingly, Ray also wrote: 'I hope shortly to be able to give you an account of the generation of some of those insects, which have been thought to be spontaneous' – presumably, the tiny flies and wasps that he and Willughby watched emerging from caterpillars.

It was actually Willughby who wrote to the Royal Society in 1671 relating how he has observed 'anomalous production' in a 'great many sorts of caterpillars'. He also says that he agrees with his friend Martin Lister's view that these maggots are the offspring of ichneumon flies. Willughby then adds that ichneumons (by which he means parasitoids in general) exhibit a wide variety of lifestyles: 'Some breed, as bees do, laying an egg, which produces a maggot, which they feed till it comes to its full growth; others, as we guess, thrust their eggs into plants, the bodies of living caterpillars, maggots &c'.[12]

Willughby was clearly close to confirming the parasitoid explanation and demolishing the notion of spontaneous generation, but the ingenious and ruthlessly competitive Lister beat him to it. Lister's priority, however, is technical only, since he had no more evidence than Willughby that parasitoids laid their eggs in other insects. Indeed, it is clear that Willughby, more cautious and much less competitive, was still seeking experimental confirmation when he says that while he subscribes to Lister's view he 'cannot yet demonstrate it'.[13]

A year or two younger than Willughby, Martin Lister had graduated from St John's, Cambridge, trained as a physician in Montpellier, and returned to England to practise in York. Brilliant, with a wide range of interests, including natural history, Lister had written to the Royal Society earlier in 1671 saying that on dissecting some large flies he had found two bags of 'live white worms of a long and round shape with black heads', and wondering whether these were the offspring of ichneumons.[14] He also thought that ichneumons fed on spiders' eggs: an error on which Willughby corrected him, pointing out that Lister had mistaken the silken cocoons spun by ichneumon larvae for the silk-enclosed clusters of spider eggs.[15]

When Ray came to write Willughby's history of insects many years later, he made a more definitive statement: 'I think the ichneumon wasps prick these caterpillars with the hollow tube of their ovipositor and insert their eggs into their bodies.'[16] Others had reached a similar conclusion, including two Dutch biologists, Jan Swammerdam, who also studied the parasitoids of white butterfly caterpillars in the 1670s, and Antoni van Leeuwenhoek – pioneer microscopist and discoverer of spermatozoa – who described the parasitoids of aphids. However, contrary to Willughby's earlier comment that silkworm larvae never exhibited anomalous generation, the life cycle of their parasitoids (tachinid flies) had been known to the Chinese silk-producers since the eleventh century.[17]

The existence of parasitoids could have spelled difficulties for the deeply religious Ray: why would an all-wise and benevolent God create something as beautiful as a white butterfly only to have it destroyed so diabolically by a parasitoid? Intriguingly, this does not seem to have occurred to Ray, and it was only in the nineteenth century that the issue of 'non-morality' of nature emerged. Although Ray recognised that animals such as dogs could feel pain, I suspect that when it came to insects Ray's views – consciously or unconsciously – were similar to those of Darwin's Catholic arch-critic St George Mivart, that they were incapable of feeling pain. For Darwin, the existence of parasitoids reinforced the efficacy of natural selection over design by a deity:

> I own that I cannot see as plainly as others do, and as I should wish to do, evidence of design and beneficence on all sides of us. There seems to me too much misery in the world. I cannot persuade myself that a beneficent and omnipotent God would have designedly created the Ichneumonidae with the express intention of their feeding within the living bodies of Caterpillars.[18]

The studies of parasitoids were only one part of Willughby and Ray's entomological endeavours. At a meeting of the Royal Society on 5 May 1670, the physician John King[19] 'produced some willow wood, containing worms wrapt up in leaves, and lodged in several channels made by themselves', which he had received from Sir John Barnard from Northampton. There being three pieces of wood, 'one was delivered to Mr Hooke for the [Royal Society's] repository, the other to Mr Willughby, and the third was kept by Mr King to observe what insect it would produce'.[20]

King describes the leaf-wrapped packages as being like 'cartrages of powder, wherewith pistols are wont to be charged', each about one inch long, composed of twelve, fourteen or sixteen pieces of leaf, and 'put one after another into a bore made in the wood, fit for their reception'.[21]

Seriously intrigued by the specimen he received, Willughby then travelled with John Ray to Astrop some twenty miles south-west of Northampton in August 1670 to see some of the 'cartrages' for themselves. The newly discovered spa at Astrop also allowed Willughby to take the waters there in the hope of curing his recurrent fevers. A Mr Snell showed them some of the 'cartrages' and Ray identified the outer covering as rose leaves, recalling 'that this very spring a worthy friend of his, Mr Francis Jessop brought him a rose leaf out of which himself saw a bee bite such a piece, and fly away with it in her mouth'.[22]

Announcing that he thinks he has revealed the 'whole mystery' of the bees, Willughby began his letter to the Royal Society with a description of how the 'cartrages' are orientated along the grain of the wood; that they contain pap or batter 'of the consistence of a jelly, or sometimes thicker; of a middle colour between syrup of violets and the conserve of red roses; of an avid taste, and unpleasant smell'; and that there is one bee larva per 'cartrage'. We now know that the malodorous pap or batter is a semi-liquid mix of pollen and nectar on which the developing larva feeds as it grows.

Several bees emerged from the original piece of wood sent to Willughby, and he describes them as being shorter and thicker than honey bees, 'but the surest mark to distinguish them is that the forcipes [jaws] or teeth of them are bigger, broader and stronger'. He also included a simple drawing – but not of the bee itself – to complement his letter.

Almost a year later, in July 1671, Willughby reported again to the Royal Society, that the 'cartrages' he had collected in 1670 were hatching and 'do now almost every day afford me a bee'. He adds excitedly that he could hear the bees 'gnawing out their way before I see them'. The arrangement in which the young ones, adequately provisioned, develop into adults, 'designed to lay all winter' before emerging the next summer, Willughby calls an admirable 'contrivance of God and Nature'.[23]

The bees that King, Ray and Willughby observed are solitary, leaf-cutter bees, and Willughby's is the first description of their life cycle. The species was subsequently named Willughby's leaf-cutter bee, with the scientific name *Megachile willughbiella*.

ↆↆↆ

John Wilkins's book, *An Essay towards a Real Character and a Philosophical Language*, was published in 1668. Included within its hundreds of pages were classifications of plants by Ray, and of birds and fish – of which more later – provided by Willughby and Ray. It also included Willughy's classification of insects ingeniously structured around 'generation' (reproduction and development), with the two major divisions being those insects whose offspring hatched from eggs as miniature adults, and those that went through

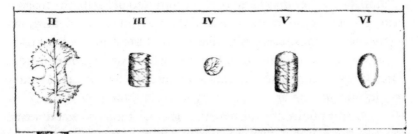

Francis Willughby's sketches relating to his leaf-cutter bees. II a rose leaf with segments removed by the bees; III a 'cartrage'; IV an individual rose-leaf piece unwrapped. The original caption reads: 'II represents the leaf, out of which a long piece, as III, and a round piece, as IV, were bitten. V shews the cartrage itself, and VI the theca [pupal case].

a succession of different life stages – caterpillar, pupa, chrysalis – before finally emerging as a winged adult.

There had been two notable previous attempts at insect classification, one by Aldrovandi in 1602, who arranged insects according to whether they were aquatic or terrestrial, with further subdivisions based on morphological characters such as the presence or absence of wings and legs. If the idea of a legless insect seems odd, remember that in the 1600s 'insects' included worms and molluscs.

The other classification was by Thomas Muffet, also known as Moffett or Moufet, but it was not published until 1634, long after his death. Muffet had been a student at Trinity College in Cambridge in 1569, going on to train as a physician in Basel. Travelling during the 1580s, he studied the anatomy of silkworms in Italy and returned to settle in London in 1584 where he and Thomas Penny, whom he had met as an undergraduate, began working together on a book of insects.[24] In the 1590s, Muffet produced a book on nutrition entitled *Health's Improvement*, published posthumously in 1655, containing a list of 100 English birds classified according to their culinary and nutritional properties but also their own diet. Written at a time when the ornithological literature was extremely limited, Muffet's information has been considered by some to be important, but others dismissed it as being derived from Gessner and William Turner.[25]

Muffet's *The Theatre of Insects*, also posthumously published, had a history that in some ways anticipated Ray's writing up of Willughby's insect observations. It appears that Penny started the book, obtaining some of his material from Gessner, whom he had met in 1565. After Penny died in 1588, Muffet inherited the project, continuing to add information and completing it in 1597, but he too died – in 1604 – before publishing the material. The book then passed through several other hands and was eventually published by a London-based Huguenot physician, Theodore de Mayerne, in 1634. Although appearing under Muffet's name, the title page of *The Theatre of Insects* is explicit in stating that it is the work of several authors.[26] The classification in it is based on the presence or absence of wings, habitat (aquatic versus terrestrial) and the number of feet, but whether it was constructed by Penny, Muffet or someone else, we may never know.

Muffet is best remembered today not for his work on insects, birds or nutrition, but for his stepdaughter Patience, who was scared away by a spider while eating her curds and whey.[27]

Both Aldrovandi and Muffet classified insects based on what the animal looked like at any particular point in time. Willughby's innovation was to use metamorphosis – how the animal *changed* over its lifetime – in his classification. It was a stroke of genius.

The reality of embracing Wilkins's project for a universal language of science, however, came as something of a shock to Willughby and Ray, for the rules within which Wilkins wanted them to work were restrictive – to put it politely. Wilkins was 'fascinated by the harmonies of nature and the mystic sense of the significance of numbers' and imagined the arrangement of animals and plants falling neatly into pre-existing groups of nine.[28] Despite this unrealistic, not to say ludicrous constraint, both Willughby for birds, fish and insects, and Ray for plants, managed to comply with Wilkins's wishes better than one might have imagined. And despite Wilkins's gushing acknowledgements in the preface to his *Essay towards a Real Character*, they weren't happy, as Ray disclosed in a letter to Martin Lister:

> It remains for me to reveal to you and pour into your bosom something that has severely stung me. I am ashamed and tired of those botanical tables, in the composing of which the bishop of Chester [as Wilkins now was] revealed that he had employed my efforts.

He continues:

> In arranging them [the tables in Wilkins] I was compelled to follow not where nature led, but to adapt the plants to the prescribed method of the author ... What possible hope is there that this method would be perfect rather than utterly deficient and absurd?[29]

Willughby almost certainly felt the same. And while Ray's letter to Lister implies that it was all a dreadful waste of time, the
exercise probably wasn't entirely without benefit, for just as an
author profits from successive drafts of a manuscript, Wilkins's
enforced exercise in classification undoubtedly helped Willughby
think hard about how he and Ray *should* classify birds and other
organisms.

Spurred on by the success of his *Essay*, which Robert Hooke enthusiastically endorsed, saying that he wished that all scientific information could be communicated in this fashion, Wilkins began
work in 1669 on a Latin edition of his book, presumably to attract
a continental audience. He asked Willughby and Ray to now provide some new, revised tables, and it may have been in connection
with this that, in April of that year, they went to stay at the epis-
copal palace on the river at Chester, where Wilkins was now the
bishop – a position gifted to him by the king in 1668 as a reward
for his *Essay*. It was at Chester that Ray acquired an unfamiliar
duck, a common scoter, from the market and managed to purchase
a porpoise from some local fishermen, which he and Willughby
dissected. Ray's account of the porpoise dissection, published in
the Royal Society's journal in 1671, described the animal's external
morphology and internal anatomy in an extraordinarily clear and
concise way. Indeed, his account is a model of how such scientific
investigations should be reported. However, even though Ray recognised that many aspects of the porpoise's internal features were
more similar to those of quadrupeds than fish, he persisted in referring to it as a 'fish'.[30]

The other event at Chester was a bout of debilitating illness
that laid Francis low for a while. It was a foretaste of what was
to come.

With more loyalty than commonsense perhaps, both Ray
and Willughby produced revised tables – in Latin – for the new
edition of Wilkins's *Essay*. Willughby's tables were evidently a
substantial improvement on the original ones, for later, after

Willughby's death, John Ray told John Aubrey that this particular classification of insects was the best there was. Sadly, the Latin edition never saw the light of day for Wilkins died – of 'a suppression of urine' – in November 1672, and, much more significantly for the history of science, Willughby's classificatory tables were lost too.[31]

Despite some initial enthusiasm, John Wilkins's attempt to produce a universal language of science was not a success – for several reasons. The notation and symbols he proposed were as alien as Chinese characters are to an Englishman, and few, if any, had the patience or inclination to master them. Wilkins's fixation on the number nine was also difficult to defend. But most importantly, his ingenious scheme failed because – as other philosophers were quick to point out – it was logically flawed. As an ardent cleric, he assumed the story of the Bible to be true; that Adam had once possessed all knowledge, including that pertaining to natural history and classification. Wilkins similarly assumed that Noah had also enjoyed that same knowledge – how else would he have been able to ensure a pair of every species in the ark?

Wilkins's view was that Adam had somehow lost that knowledge, which like original sin, was innate and built into us, and that it was his task to recover it. If he'd known the term, he'd have said it was imprinted in our DNA. The philosopher John Locke on the other hand reckoned we were all born as a tabula rasa, a blank slate, and that knowledge is not innate, and hence is not recoverable, a view that anticipated the nature-nurture debate.[32]

Curiously, Wilkins's numerically constrained arrangement of the natural world foretold and indeed may have inspired a similar system in the early 1800s by the entomologist William Sharp Macleay, but one based on fives rather than nines. Macleay's system was rooted in the notion that everything in the natural world could be represented by interlocking circles – an idea that harked back to Pythagoras – and that there were five orders of birds, subdivided into five tribes each comprising five families. The most enthusiastic ornithological proponent of what became

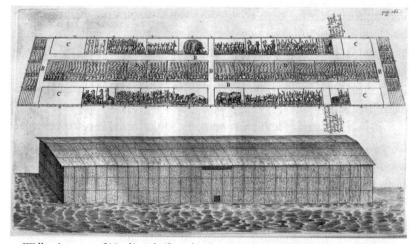

Wilkins's image of Noah's Ark (from his Essay towards a Real Character, *1668) – with a complete absence of birds, fish and insects.*

known as the 'quinary system' was William Swainson during the 1830s. Like Wilkins's arrangement based on the number nine, this too was nonsense, but for a while at least it was a roaring success, unlike Wilkins's system. The ornithologist Alfred Newton later wrote:

> The success it [the quinary system of classification] gained was doubtless due in some degree to the difficulty which most men had in comprehending it, for it was enwrapped in alluring mystery, but more to the confidence with which it was announced as being the long-looked-for key to the wonders of creation, since its promoters did not hesitate to term it the discovery of '*the* Natural System'...[33]

I should add that the enthusiastic acceptance of the quinary system by so many nineteenth-century ornithologists was also a measure of the desperation with which the true 'arrangement' (classification) of birds was sought.[34] Alfred Newton's words sparked a sense of déjà vu for me, as I thought about the subsequent announcements of the discovery of the 'true' arrangement of birds, and the way that molecular revelations are also – for many birdwatchers and ornithologists – enwrapped in alluring mystery. We always assume that what we know at any point in time is the truth, forgetting that as has been shown over and over again, in many instances it is usually only the 'truth for now'. This is the way science proceeds.[35]

In the words of one historian, the task Wilkins set himself with his project for a universal language of science was as ambitious and monumental as nuclear fusion: splitting the atom. The end result, however, was more like the 1980s notion of 'cold fusion', a nuclear process that promised clean, cheap energy, but turned out to be a very damp squib.

Wilkins's project wasn't a complete waste of time. The work he inspired Willughby and Ray to undertake resulted in classifications of birds, fish and insects that were to stand the test of time and form the basis of those used by all subsequent generations of biologists.

᚛ᛁ᚜

Across the fields and several miles from Middleton Hall lies a small copse. Francis Willughby and John Ray are following a young farmhand towards the wood, and once inside he stops beneath a giant oak and points up to the topmost branches to show them the large, flat twiggy nest he has found. All three men step back into the shade of the understorey, but with a clear line of sight to the nest. Francis holds the gun expectantly. All eyes are on the nest. They wait. After twenty minutes, in a sudden sweep of stalling wings a huge bird appears through the canopy and the gun goes off with a crack and a cloud of blue smoke. The bird falls noisily through the leafy branches and onto the ground. It staggers into an upright position and looks around in a bewildered kind of way. Willughby, Ray and the farmhand set off towards their prey, but the bird sees them and on flapping wings, but unable to take flight, runs as rapidly as a chicken through the undergrowth. Faster, the farmhand overtakes the bird, and kills it by twisting its neck. Urging him not to be too vigorous, Francis takes the limp body from him, excited by his new acquisition.

They walk back to the tree that holds the nest. The farmhand swings himself up into the lower branches and then climbs carefully upwards. He peers over the rim of the nest and announces loudly that there are two chicks. Somewhat warily in case they peck or claw at him, he grabs each in turn and throws them down onto the ground where Willughby kills them. There's no emotion in this. It is functional: the acquisition of specimens. While John Ray

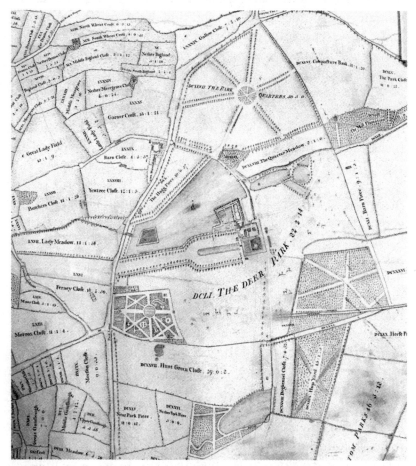

Middleton Hall and immediate surroundings in 1762. The house, with its moat (centre), was probably much as it was in Francis Willughby's day, but the avenues of trees and formal gardens were probably planted by Francis's son, Sir Thomas Willoughby, in the early 1700s

Monstrous but marvellous fake beetle purchased by Francis Willughby during his continental travels

Moorish gecko (top) and Hazel or common dormouse (bottom); paintings obtained by Francis Willughby during his continental travels (artists unknown)

Willughby's leaf-cutter bee (top), and bee 'cartrages' containing a single or egg or larva and constructed inside a hollow stem (middle) – each cartrage is separated by a mixture of chewed mix of leaves and pollen; Willughby's charr (bottom)

The extant part of Francis Willughby's egg collection from various bird species (top); egg of a snipe labelled in Willughby's handwriting (middle left); egg of a night heron *Ardea cinerea minor* (middle right), possibly with Philip Skippon's handwriting; egg of a starling ('stare'), unknown handwriting (bottom left); and part of an egg of a grey heron, possibly labelled in Skippon's hand (bottom right)

Part of Francis Willughby's collection of curiosities (top), with close-up of glass eyes, mollusc shells and insect parts (middle), and coloured minerals, mollusc shells, a glass eye and a fake beetle

Dried plant specimens from Francis Willughby's collection: hare's-tail cotton grass (*Eriophorum vaginatum*), with annotations by both Willughby and Ray

Two trays of seeds from Francis Willughby's seed cabinet or 'cabinet of curiosities'

Willughby's buzzard, otherwise known as the Eurasian honey-buzzard, which Francis Willughby recognized as distinct from the common buzzard

may once have waxed lyrical about the beauty of wildflower meadows as the root of his botanical inspiration, there's neither lyricism nor sentiment here. Willughby is pragmatic. There is no ornithology without specimens: it is as simple as that. We might recoil in revulsion at such an attitude, but for Willughby and Ray killing a bird was little different from us today identifying one through binoculars. Understanding the beginnings of ornithology requires us to recognise this, putting aside our emotions, and for the time being at least, accepting that the quest for knowledge was emotionally neutral, at least in terms of specimens.

Back at Middleton Hall, Francis lays out the three birds, the adult and the two chicks, on the deal table in his workroom. As though he is holding a baby directly in front of him, he picks up the adult bird in both hands and smooths its feathers before placing it on a pan balance to record its weight: thirty-one ounces. Ray takes note. Willughby then measures the length of the bird from the point of its bill to the tip of the tail (twenty-three inches); the span of its wings (fifty-two inches); and from the tip of the bill to the corner of its mouth (one inch and a half). The nostrils, he notes, are 'long and bending [curved]', and the irises a 'lovely bright yellow or saffron colour'.

After describing the plumage, including its ash-coloured head and mouse-dun back, he lays the bird on its back and lifting each wing in turn counts the primary and secondary feathers together – which he calls the flags – twenty-four on each side. The tail, he notes, consists of twelve feathers. Moving now to the legs he tells Ray that they are 'feathered down somewhat below the knee' (more usually known as the 'ankle', technically this is the joint between the tibiotarsus and tarsometatarsus). The legs and feet are 'short, strong, yellow … The talons long, strong, sharp and black.'

Ray writes it all down. Francis then picks up a small sharp knife, and parting the feathers on the bird's belly, exposes the bare skin and plunges the blade just below the sternum. He slices downwards, and then, using the fingers of both hands, pulls the sides apart to expose the guts and gently separates their coils from the connective tissue, commenting to Ray that the gut seems shorter than in the common buzzard they had examined recently. Sliding his fingers down the length of the slippery intestine, Francis reaches the bulging stomach. Holding it between his finger and thumb, he slices it open using the lance, only to find it stuffed with vivid green caterpillars, some of which he can identify. So distracted is he by this entomological bonus that he forgets to record the bird's sex.

Turning to the young birds, Francis barely bothers to describe their appearance – immature birds are of limited use in identification – other than noting that they are covered in white down, spotted with black; their feet are a pale yellow and their bill is white. Inside the stomach of the first one he finds the remains of several frogs, and in the other, he shows Ray in amazement, are 'two lizards, entire, with their heads lying towards the birds mouth, as if they sought to creep out'.

Francis washes his bloody hands, and takes stock. Turning to the notes he made previously when describing a common buzzard, he starts to list the differences between the two species. This one, known locally as a honey-buzzard, has: (i) a longer tail; (ii) an ash-coloured head; (iii) the irises of the eyes are yellow; (iv) thicker and shorter feet; (v) dark barring on the wings and tail, which are about three inches broad. 'Have we seen this bird before?' he asks Ray.[36]

Willughby enjoyed several years of pre-marital freedom during which time he continued to work on his ornithology and other projects. In June 1667, Ray wrote to Martin Lister telling him how he and Willughby had spent the previous winter arranging Willughby's collections of birds, fishes, 'shells, stones and other fossils, dried plants and coines'.[37]

Arranging specimens and creating classifications were perfect activities for cold winter nights, but the summer of 1667 saw them out and about again, journeying to the West Country and allowing Willughby to witness what he had missed because of illness in 1662. Unusually for Ray, the 1667 excursion isn't well documented, but we know that they rode through the counties of Worcester, Gloucester,

Somerset, Devon and Cornwall, visiting the botanically rich Lizard peninsula, and reaching Land's End on 17 August. From Ray's few notes, the trip appears to have been mainly botanical, although it is hard to imagine Willughby passing up the opportunity to seek new birds or to find fish in the Cornish coastal villages. They came back by way of Hampshire, reaching London on 13 September 1667, from where Willughby returned to Middleton and Ray to Notley, where it was his turn to endure a bout of serious illness.[38] After he had recovered, and been elected a Fellow of the Royal Society on 7 November 1667, Ray, or 'John Wray' as he signed himself in the Society's Charter Book, returned to Middleton to continue working with Willughby.

A year after returning from his continental expedition and with no sign of him thinking about marriage, Francis's family and friends began to apply some gentle but persistent pressure. His daughter Cassandra recounts how, on going through her father's papers many years after his death, she found several letters from relatives and friends suggesting potential partners. Well off and well connected socially, Willughby was a fine catch and families with eligible daughters were undoubtedly keen to arrange introductions. The first of the letters is one dated 4 October 1666 from a family friend, Thomas Alured, recommending 'a Sussex lady', whose identity for us remains unknown. Alured's next letter suggests Elizabeth, the daughter of a distant relative, Lord Willoughby of Parham, governor of Barbados (recently 'shipwracked' and drowned while trying to retake St Kitts in the Caribbean), being 'fit for his [Francis's] humour, and [with] proportionable fortune'. The next possibility Alured suggests is Lady Springet's daughter (possibly the widow of Sir Herbert Springate of Sussex), Barbara, whom he describes as 'a lady beyond exception, and that both mother and daughter were so in love with his [Francis's] character that his terms may be fully answered' – presumably a reference to any marriage settlement.

Alured is nothing if not persistent, and his next suggestion is the eldest daughter of Lady Pile, whose late husband was Sir Francis Pile of Compton Beauchamp, Berkshire. It isn't clear which of Pile's three daughters Alured is referring to, but he says she 'would have ten thousand pounds, besides perquisites to a lady of her quality, and that as much as the modesty and honour of that sex would permit, he was promised a welcome'. Fifth, and final, as far as we know among Alured's recommendations, was 'a lady of Lord Fairfax's family'. Lord Fairfax, otherwise known as 'Tom the Black' for his dark eyes and looks, had fought for the Parliamentarians during the Civil War, which, given Willughby's Royalist sympathies, smacks of desperation on Alured's part in suggesting this unnamed lady.

It is difficult to know whether Francis was amused, wearied or intrigued by Alured's parade of possible partners. We do not know whether he dismissed his suggestions out of hand, or whether he did so after being introduced to the ladies in question. But Alured wasn't the only one urging Willughby to marry. In February 1667, Philip, Lord Wharton, asked John Wilkins to propose one of Wharton's own daughters as a possible partner. Sir Thomas Wendy was also increasingly keen to see Willughby married. In one letter, clearly somewhat despairing of his brother-in-law's reluctance to commit himself, Wendy states that he is making 'one attempt more upon the old, but hitherto unsuccessful errand' – this time suggesting the daughter of Sir Edward Turner, Speaker of the House of Commons. We do not know the daughter's name, but in an effusive effort to find Willughby a partner, Sir Thomas described her as someone Francis 'might be abundantly happy with', adding as an additional inducement that 'She has been educated with very great care'. Not only that, she was 'of a vertuous inclination, sweet disposition, and serious composure of mind'. Moreover, having once been in her company, Sir Thomas judged her to be 'of acceptable conversation, of a very good stature, about twenty years of age, tho not curious [i.e. not exceptional in appearance] yet cumely', to which he added that her portion of any marriage settlement would be 'seven thousand pounds paid down'.[39]

All to no avail.

Sir Thomas persisted, and his next suggestion was Emma, the younger of Sir Henry Barnard's two daughters. Francis and Emma were introduced sometime during 1667, but details of their courtship are lacking. They were married in early January 1668 at her home in Bridgnorth, Shropshire: Francis was thirty-two and she was twenty-two or twenty-three. The marriage settlement, the deal to safeguard the wealth of both sides, which was overseen by their respective parents, was painfully protracted and one in which the well-educated Emma took an active part. Francis, too, became increasingly frustrated by the time taken to tie up the deal, which wasn't resolved until three years after the wedding.

The couple went to live at Middleton Hall where Willughby's mother, the formidable Lady Cassandra, continued to preside. The marriage seems to have been a great success, with deep affection on both parts. Francis's sister Katherine, whose own marriage was difficult and who, with her young son, spent much of her time with her mother at Middleton, thanked Emma for the happiness she had brought her brother. Lettice Wendy, Francis's other sister, wrote to Emma soon after the marriage and before meeting her, to say how pleased she was that her brother had 'pitched on one soe suteing him'. Katherine and Lettice clearly approved. Lady Cassandra too must have been delighted, and relieved, that her unconventional son had finally started to settle down.[40]

Willughby and Ray continued to work on their bird material and their friends continued to send them notes and specimens. Emma's father wrote to Francis in 1668 reporting the capture of an unusual bird that he hoped would end up in Willughby's hands. 'It is almost as big as a cuckow, long wings as a Martin, speckled like a Woodcock, a sharp little bill or beak, the eyes standing backwards as big as an Owles, with long hairs on each side of the beak like a Ratt, with some white feathers on each wing. The like hath not been seen here by any of the oldest faulkners [presumably falconers].' The bird was a nightjar and in Shropshire, where it was captured, the species is still considered rare.[41]

Willughby and Ray's friend, Francis Jessop – a student friend from Trinity College and now a wealthy mill owner as well as a mathematician and naturalist, living in Sheffield on the edge of the Derbyshire Peak District – sent Willughby the 'cases', or bodies, of various birds.[42] The city of Sheffield, then comprising some 5,000 souls, lies up against two very different landmasses on its western border: the Dark Peak comprising millstone grit, and to the south the softer limestone of the White Peak, each with very different bird populations. On one occasion, Jessop sent Willughby several goldcrests – known then as the golden-crowned wren or copped wren – he had captured at his wife's family home at Highlow Hall near Hathersage some eleven miles from Sheffield. The goldcrest was duly described in the *Ornithology*, but with no acknowledgement of the similar but rarer firecrest. Interestingly, Conrad Gessner did the opposite in his 1555 book, describing what is clearly the firecrest, but with no mention of the goldcrest, and presuming, as Willughby had done, that there was but a single species in the genus *Regulus*.[43]

From the 'mountains of the Peak of Derbyshire', Jessop sent specimens of the twite (known then as the mountain linnet), a warbler which he called a 'pettychaps', a male and female brambling, ring ouzel, dipper, long-eared owl, and, from an inland body of water near Sheffield, a puffin and a common scoter. The puffin must have been a young bird blown far off its usual marine habitat. The scoter, also a misplaced marine species, preceded the one that Ray was to acquire and describe (as he thought, for the first time) from Chester Market in 1669.

In the autumn of 1668, Jessop obtained the skin of a 'great bird' that he was told was a 'scarfe', but believed would 'prove [to be] a bernicle', and sent a description of it to Willughby. A 'scarfe' was the Lancashire name for a shag or cormorant, or possibly a barnacle goose – a species that also occurs on the Lancashire coast in winter. It is difficult to imagine anyone confusing a cormorant or shag with a goose, so the bird's identity remains unclear.

Another mysterious specimen that Jessop sent to Willughby was a 'black legged linnet'. 'A friend of mine,' he wrote, 'kept it in a cage till it dy'd; and so it lay neglected, until I found it by chance dried, as you have it.' Later, in the *Ornithology*, Ray wrote: 'Mr Willughby ascribes to the feet of this bird an obscure dusky or blarkish [blackish?] colour', which clearly contrasted with the description in Olina's book, *Uccelliera*, in which he states that the legs are a 'middle colour between flesh-colour and white'.

Ray – who appears to have had no personal experience of a lin-net's legs – wondered whether there might be differences in the leg colour between male and female linnets, or between birds of different ages, completing his account of the species by saying 'Mr Jessop sent us a linnet … with feet perfectly black, but that was extraordinary.'[44] I suspect Ray was right and that Jessop's bird was an ordinary linnet whose legs had turned black either with age or after it died.

From the Dark Peak, Jessop obtained two 'moorgame' for Willughby and Ray. In his letter accompanying the box of birds he wrote: 'There is a cock and a hen; the cock is pretty perfect, but the hen hath a wing shot off.' Uncertain as to what they were, Jessop continued:

> The moor-cock is certainly none of the *Gallina Corylorum* [hazel grouse]; and whether it be the *Grygallus* [female capercaillie or female blackgrouse] which Gessner describes, I also doubt, having compared these with both his cut [image] and description: it agrees with it in many particulars, but differs from it in some. The feet are not like those of the *Urogallus minor* [black grouse] but nearer resembling those of the *Lagopus* [ptarmigan] being feather'd all over.[45]

Jessop hoped Willughby and Ray could establish the identity of the moorgame, but they were confused too, and on 10 September 1668, Ray, while staying with Jessop in Sheffield, wrote to Martin Lister:

> Of birds only four or five species were found by me that I had not seen before, to wit the *Grygallus major* of Gesner, which the Italians call a Francolin; it is common on heather-clad mountains; sportsmen and countryfolk call it the Red Moorgame. I am well aware that Gesner thinks that the Italians' Francolin is called the Hazelhen. I think this bird is the same as that whose picture Thomas Crew showed us at Montpellier whose French name I have forgotten.[46]

This illustrates just how difficult it could be to distinguish similar species and to match written accounts with images or real specimens. Unless sufficiently accurate and detailed, written descriptions can be extraordinarily difficult to interpret, which is undoubtedly why Willughby was so meticulous in his notes. Illustrations, one had to assume, were truthful representations of nature, but as Ray's written comments on Thomas Crew's painting referred to above indicate, one couldn't always take that for granted. Examining specimens, preferably alive or recently dead birds, or dried study skins, is best, but even they show variation between individuals, age classes and the sexes.

The red moorgame is the red grouse, common – as Jessop says – on Sheffield's Dark Peak. Gessner's bird 'common on heather-clad mountains', presumably in the Alps, must have been the rock ptarmigan (which Willughby had seen in Italian bird markets), and the bird illustrated by Sir Thomas Crew, was, as we have seen, something completely different: a pin-tailed sandgrouse. Remarkable as it may seem for what is now a rather familiar bird, Willughby and Ray's may have been the first description of the red grouse. They could not then have known that this bird is endemic to Britain and today is considered a distinct form of the willow ptarmigan (which is widespread across the northern hemisphere), but differs from other populations by not turning white in winter.

In the same letter, Jessop described how he had been working hard all day to solve a geometrical problem that Willughby had clearly asked him about. The letter is a mere fragment of what is obviously a longer conversation, so it is unclear just what the issue was, but reading between lines I think it may have been the old chestnut of squaring the circle.[47]

Willughby and Emma's first son, Francis, was born on 13 September 1668, eight months after their marriage and 'above a month before his time'. He apparently had 'no nails grown' and was so 'tender and little' that he needed very special care during his first few months.

It isn't uncommon for premature babies to have very small, under-developed nails on their fingers and toes. Like all parents, Emma and Francis obviously thought the world of their new child – 'dear Jewel Frank' – and, as he developed, they also believed him to possess uncommon intelligence. Emma recounted an event that to her, at least, confirmed the child's extraordinary cognitive abilities. She had given young Francis, then less than a year old, the bodkin (pin) from her hair to play with. Unseen by either his mother or any servant, he pushed it through a crack in the floorboards. A full year later when the bodkin was mentioned, he showed his mother the crack and when the floorboard was lifted, there was the bodkin.

Early in 1668, when there were no insects to investigate, Willughby and Ray busied themselves with investigations of the flow of sap in trees. If we were to judge simply from their results – published in 1669 – we might have assumed that these particular studies began only in the previous year, and in response to a set of 'Queries concerning vegetation, especially the motion of the juyces of vegetables', published by the Royal Society in January 1668. But in fact we now know, from hastily written notes in Francis's commonplace book, that he began these experiments almost three years earlier, in March 1665, and he did so, moreover, on his own, for John Ray was still on the continent at this time.

Willughby's notes from 1665 provide another insight into the way he approached this study and reported his findings:

Towards ye End of March <A[nn]o 1665> there was several Transverse incisions made in Birch under ye Bowes a little deeper then ye barke. Ye wounds were kept open with little Wedges of wood, and ye liquor was Conveied into bottles by Filtres of Flannel.

Ye principall observations were,

1. That in ye night (or in ye day when there was very cold blasts) scarce any liquor came.
2. They allwaies dropped Fastest in ye morning ye Sap being then most greedily sucked up.

3. Ye greatest Bowes and those that were nearest ye roots afforded most.

4. They continued dropping at ye same wounds neare a Fortnight.

5. When they begun to give over a thick white substance like Fleame stuck to ye Flannels and settled at ye Bottomes of ye Bottles. This white settlement is doubtlesse ye matter of ye white wood. Q[uaere]: Whither ye settling of ye sap bee not allwaies of ye colour of ye wood.

6. It tasted very pleasantly allmost like water and sugar, but when ye weather grew Hot, it quickly turned and grew soure.

7. Some of this was boiled with sugar &c. as Mr Evelin Directs.[48]

To paraphrase: by making cuts in birch trees he found that the flow of sap was least during the night (or if the day was cold); at other times the flow was greatest in the morning and especially vigorous from large branches and those nearest the roots. The sap continued to flow from cuts for almost two weeks. As the flow dried up, a phlegm-like substance appeared, which he assumed was part of the 'wood'. Willughby says that 'it' tasted like sugar water, but it isn't clear whether he is referring to ordinary sap or the phlegm-like sap just mentioned. I suspect he's referring to ordinary sap since he then says he boiled some, as 'Mr Evelin [John Evelyn]' suggested. Boiling the sap collected from trees in spring has been practised by native peoples for a long time and is how maple syrup is made.

The other significant aspect of Willughby's notes is the inclusion of three queries that Francis asked himself: whether sap was always the same colour as the wood; whether sap comes only from the bark; and 'whether ye sap descend and circular [presumably meaning "circulate"] or evaporate by perspiration'.

William Harvey's discovery in 1628 of the circulation of the blood – ingeniously demonstrated using tourniquets and simultaneously establishing the existence of valves in veins to prevent any backflow – raised the question of whether a similar circulation occurred in plants. Previously, it had been assumed that plants had no 'sensitive soul', an idea promoted by Aristotle and Theophrastus in the fourth century BC. However, the fact that the sensitive plant *Mimosa pudica* from tropical America, recently discovered in Willughby's day, has leaflets that droop dramatically on being touched and reopen minutes later, suggested that it was far from insensitive. In fact, this plant, whose extraordinary and rapid movements continue to fascinate us, was the starting point for studies of plant physiology.

Understanding how *Mimosa's* leaf movements occurred was one of the first tasks allotted to the newly formed Royal Society by Charles II. In response, the Society established a committee in 1665 to create a list of botanical questions for its members. Robert Hooke made the study of *Mimosa* his own, but the Society's list – eventually published in January 1668 – also included queries relating to the rising of sap in trees: 'In tapping, cutting or boring of any tree, whether the juyce, that vents at it, comes from above or below … What side of the tree affords most sap?'; 'whether the sap comes more copiously at one time of the day or night, than another?'[49]

Willughby embraced the study of sap with the same enthusiasm and ingenuity as he did his other research. The questions and the methods that he – and later Ray – employed might seem naive to us now, but these were some of the first investigations into what we now call plant physiology. The original questions, and Willughby's report with Ray in 1669, prompted others, including their relentlessly inquisitive and academically ambitious friend Martin Lister, to undertake more research on the same topic.[50] Curiously perhaps, the analogy with human blood-flow resulted in the conclusion that sap circulated rather than flowed directionally through the vessels of plants. This erroneous view persisted until 1727 when Stephen Hales – whose love of botany was inspired by Ray's writings – disproved it. The remarkable way in which trees, including 300-foot redwoods, Douglas fir and the mountain ash (a kind of eucalyptus) transport sap from their roots to the uppermost branches against gravity with no obvious pump, was finally revealed in the twentieth century. The mechanism, known as 'cohesion tension', is based on there being a continuous column of water between root and leaf and the surface tension caused by evaporation from the leaf surface pulling the sap upwards through the entire height of the plant.[51]

The years at Middleton after their continental journey saw Willughby maturing as a scientist and creating a niche as a natural historian within the Royal Society. Those years saw him also as a family man, with his daughter, Cassandra, born on 23 April 1670, and a second son, Thomas, on 9 April 1672. Soon after Cassandra's birth, Lettice Wendy wrote to congratulate Emma, noting, somewhat disapprovingly, that her brother was away again, but finished by saying, 'well, you know what he's like'.[52]

Increasingly, Willughby sent his observations, many of them made with Ray, to Henry Oldenburg, the Royal Society's correspondence secretary. The result was that most were duly published in the Society's *Philosophical Transactions*.

One of Charles II's achievements soon after gaining the throne was to establish the General Post Office in 1660, whose rapidly evolving postal service became a boon to those Royal Society members, such as Willughby, who lived too far from the capital to attend its weekly meetings. Although letters had been important for scholars for well over a century, Francis could now write a letter reporting his findings to Oldenburg, knowing that it would be read to those members present at the Society's next meeting, and then receive feedback, usually by return of post.

Creating a letter in the 1660s involved a kind of literary origami, which, since the invention of the envelope, is a skill long since lost. Taking a single sheet of paper orientated in what we would call the 'landscape' position, Francis – and many others, it seems – folded it vertically down the middle. He then wrote on both the first and second sides, as one would normally write on these now 'portrait'-orientated half-pages. After completing the second page he turned the paper through 90 degrees to continue writing – typically – half a page, at right angles to the first two pages. He then folded the

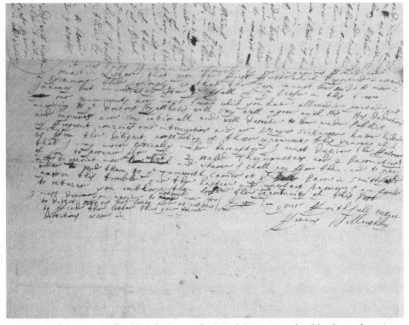

One of Francis Willughby's letters to the Royal Society on the 'bleeding of trees'
(the flow of sap), dated 12 March 1669.

paper inwards on each side to create three equal-sized blank panels. Turning the folded paper over, in the middle of the panel he (or sometimes someone else) wrote: 'For Mr Henry Oldenburg secretary to the Royal Society at his house in the Pal Mal, London'. Often as an afterthought, Francis continued to add information on the other two panels. Finally, the letter was turned over and the two flaps were sealed with a blob of wax, such that the letter formed its own envelope.

Opening Francis's letters today, they look – there's no polite way of putting this – a bit of a mess, an impression not helped by his hasty scrawl using every bit of available space. But this was Francis's way of doing things, and it reflects his energy and eagerness to get things done. Even his formal letters appear to be rapidly written.

Francis's letters would have been handed to a boy who walked to the post house or receiving station (usually an inn) at Tamworth six miles away. From there a man on a horse would carry Willughby's letter, along with any others, down the post road – the Chester Road (now the A5) – in eight- to sixteen-mile stages, changing horses each time, towards London some eighty miles distant. The Royal Society, in turn, had someone who visited the London receiving station each day to collect the incoming post and deliver the outgoing letters.

Reading through Willughby's letters of discovery in the Royal Society library, I have a strong sense of a man finding his footing and warming to the task of reporting his findings. The free exchange of information and the informality with which he and Oldenburg corresponded is endearing. Such familiarity occurs occasionally today when author and journal editor know each other, but in the main, the process is rigidly objective, automated and utterly impersonal. Oldenburg, however, deliberately and skilfully developed relationships with his correspondents, encouraging, cajoling and orchestrating their efforts for the sake of science. Deeply committed to the Society and his position within it, Oldenburg worked his socks off, writing 250 to 300 letters each year, commenting once to Robert Boyle: 'I am sure, no man imagines what store of papers and writtings passe to and from me in a week.' Oldenburg, whose personal circumstances are little known, was the glue that held the Royal Society together during its first decade, dexterously constructing a vast web of practitioners with the explicit aim of fulfilling the Society's objective to scrutinise the whole of nature.[53]

On 19 March 1670, Oldenburg wrote to Francis Willughby to tell him of a letter he had received from Israel Tonge, another Fellow, that included two noteworthy bits of information: an account of the circulation of sap in trees, and a duel between a spider and a toad.

Intrigued, Willughby replied:

I shall impatiently desire very particular directions for the duell between the spider and the toad, and a good description of the spider [presumably, if he was to identify it]: I will use all possible diligence in any necessary apparatus, it being a thing so strangely improbable, that I shall scarcely believe my owne eyes.

Oldenburg must have asked Tonge for further information because on 16 June, Tonge sent a more detailed if somewhat incoherent account. It is immediately apparent that he had not witnessed the duel himself, but in an effort to acquire more information, and to his credit, he had visited the site – the garden of the White Lyon alehouse at Marden, near Maidstone in Kent – where the duel was said to have taken place. After asking around, he located the sole surviving witness, Elias Rolfe, who told him that the incident occurred in 'ye time of ye late warre', presumably the Civil War, which ended in 1651 – so about twenty years previously, and when Elias was ten or eleven years old.

He and several others of the household:

> … heard a strange squeaking noise, and going forth to fetch some billets from ye woodpile hee perceived, yt ye noise, they had heard in the brewhouse, was made by a toad in a fight with a spider. The manner of their fight was like yt of 2 fighting cocks, sometimes approaching sometimes retreating. The manner by wch ye toad did labor to defend himselfe, was by spitting wch he cast from him a foot or 2, and ye spider avoided it by leaping up on high, so sometimes she would leap above a foot right up from ye ground. This skirmish continued for ye space of an hour & more, till at last ye spider got upon the toad & killed him.

Elias added: 'The spider was a large one of ye bignesse of a child's fingers end and white bag'd, had a white dun list down her sad coloured back.'[54]

Remarkably, Tonge found other people in the same village who had apparently witnessed separate spider-toad incidents. One described how the spider jumped onto the back of the toad and 'fastens her teeth into his neck in ye place where his head is joined to his body' and that the spider let go when the toad squeaked.

Finally, Tonge asked Oldenburg: 'But herein I must request Mr Willoughby's information of whom I long since was informed, that he is curious in such observations.' Willughby certainly was curious about such things, but unfortunately there's no record of him expressing an opinion.[55]

My feeling is that the spider-toad contest was a folk tale, elaborated and multiplied by time and false memory. That *three* such toad-spider duels had occurred in the same area seems unlikely, especially since none of the several spider experts I consulted had ever heard of spiders attacking toads.[56]

᛭ᛏ᛭

As Francis Willughby is best known for his contributions to natural history, it comes as something of a surprise to discover that he also wrote a book on games. It certainly was an eye-opener when his manuscript was found among the papers in the Middleton Collection during the 1970s. Prior to this discovery, no one ever imagined that Willughby's intellectual pursuits ranged so widely. Neither John Ray nor William Derham mention his interest in games, presumably because it had little bearing on natural history. Indeed, it is probably precisely because of this lack of interest that Willughby's study of games survived, lying untouched and unread for over 300 years among the family papers.

Mary Welch, archivist for several decades at Nottingham University Library, was the first to handle the Middleton Collection, and it was she who, in adding the material to the university catalogue, sometime in the 1950s or 1960s first referred to the volume as the *Book of Games*. Her description of it is as 'a large folio, parchment-bound volume containing notes which he [Willughby] compiled on card games ("nodde cribbage", "ruffe and trump"), ball games (football, stool-ball), other games requiring special equipment (shuttle-cock, "scotch hopper", bowls, and "ten pegs or 9 pegs", shovel board, and so on), and simple children's games'.[57]

Although Welch christened it the *Book of Games*, Willughby himself would probably have called it a book of 'plaies' rather than games. For Welch the manuscript was one of several, and while of considerable interest, she attached no particular importance to it, assuming merely that Willughby had taken part in some of the sports he describes.

It wasn't until Dorothy Johnston, Welch's successor, looked at it again in the early 1990s that the manuscript's true significance emerged. Dorothy discussed the manuscript with her partner, David Cram, a linguist, and he realised that it was very different from anything else previously written on games, and much more than just a list of sports or 'plaies'. Not only that, the manuscript is one of only a handful of documents in which we can hear Willughby's voice directly.

Prior to the mid-1600s, all accounts of games had been written *by* gamesters *for* gamesters. Willughby's interest was completely different. While cribbage, football and scotch-hopper might seem somewhat remote from bitterns, burbots and bees, we perhaps shouldn't be too surprised that Willughby's interests were so broad. Francis Bacon's philosophy, which served as a beacon for members of the Royal Society, encouraged 'histories of all kinds' – that is, first-hand investigations of any topics that could contribute to a better understanding of natural and man-made phenomena.

Willughby's interest in 'plaies' was probably piqued during his travels in 1662 with Ray and Skippon as they watched local people participating in various games and sports.[58] It is unlikely that Willughby immediately planned a systematic study of games, but with his intellectual curiosity aroused he started to document what he saw on this and subsequent travels.

There is a somewhat cryptic reference to Willughby's interest in this topic by his daughter Cassandra, who in her memoir wrote: 'There are in the library at Wollaton many manuscripts which were written by my father ... one which shews the chances of most games.' This is probably not the *Book of Games*, but a companion volume referred to by Willughby himself as a *Book of Dice*, now sadly lost, but almost certainly dealing with the likelihood of rolling certain numbers with one or more dice. Willughby's fascination with chance, which mathematicians now call 'probability', was probably encouraged by the fact that great mathematicians such as Christiaan Huygens and Gottfried Leibniz had pointed out the value of studying games in relation to the 'doctrine of chances'.[59]

For Francis, 'games' or 'plaies' covered several different topics, from those with cards or dice, to ball games such as tennis, rural games, cockfighting, 'bare [sic] baiting', courtly word games and vulgar word games, some of which were called 'Selling of Bargains'. Willughby notes that 'All bargaines are either obscene or nastie: A bids B repeat Oxe Ball so manie times in a breath. B repeating fast saies, Ballox.'[60] In discussing cock fighting, Willughby repeats the advice that to produce the best fighting birds, fighting cocks need to be reared under ravens – possibly easier then than now, but still something of a logistical challenge!

The *Book of Games* also includes those played by children, and, curiously, several pages of it are in a child's hand, edited and corrected by Willughby himself. It would be nice to think that this was the hand of Willughby's son Francis, but he was only four when his father died so this seems unlikely. Another member of the family may have written it. However, Willughby did observe and record his son Francis's games, noting the toys and objects he first played with. Reading this volume, I was reminded of the fact that two centuries later Charles Darwin did much the same, observing the behaviour of his own children and documenting the way their facial expressions and emotions developed over time.

Another intriguing aspect of Willughby's study of games was his description of motion – including the spin on rebounding tennis

Seventeenth-century dice of the kind Francis Willughby would have seen and might have played with.

balls. It was an area that lent itself to mathematical analysis, which as we have seen was one of Francis's areas of expertise, and in 1669 he attempted to re-ignite a debate about the theory of motion, sending sophisticated analyses to Henry Oldenburg, but failing to enter the fray with the Royal Society's other mathematicians. It was a closed shop and Francis wasn't part of the tight circle of mathematical protagonists. He lived too far away to attend all the Royal Society meetings; moreover, depending as he did on published versions of the debate, he was perpetually out of date.

Previously, however, at his second meeting at the Royal Society in 8 November 1662, Francis had demonstrated his mathematical ability by presenting the solution to a long-standing geometric problem: the optimal pattern of planting fruit trees to ensure that they all received maximum light. The puzzle was probably similar to what is known today as 'circle packing': deciding how to fit the maximum number of equal-sized circles into a square of a particular size. The Royal Society later reported that 'Mr Willughby produced his demonstration to prove that the same area of ground planted with trees after a quincundiall figure will hold more trees placed at the same distance from one another, than the Square, in the proportion of 8 to 7.'[61] 'Quincundiall' here means 'quincuncial' and refers to the quincunx pattern, like the five pips on that particular side of a dice. I wonder whether Francis was inspired to tackle this problem by Thomas Browne's 1658 book *The Garden of Cyrus* – a bizarre and difficult volume that may possibly have made more sense to Willughby than it does to most modern readers. Some have even wondered whether Browne composed it while in a state of 'altered consciousness', due to its copious examples of God's quincundial geometry: lattices, the figure 'X', the number five – all presented as evidence of His wisdom.[62]

Using almost identical methods to those he employed with birds, fish and insects, Willughby constructed a hierarchical classification of games using bifurcating and trifurcating arrangements, as is apparent from the figure below.[63]

At one level then, the *Book of Games* was an intellectual exercise that allowed Willughby to utilise and develop his interest and

A. Wilkins's classification

Recreation

of the mind
- depending on chance only
- depending on chance and skill
- depending on skill only

of the body
- in respect of the whole [football, bowling, dancing, wrestling etc.]
- in respect of the eye or the ear [including theatre and music]

B. Willughby's classification

'Plaies' that exercise

the wit
- those that have nothing of chance
- those that altogether depend on fortune
- those that have both art and skill

the body [including Tennis, Stowball, etc.]

Willughby's classification of games compared with John Wilkins's classification (upper section).

knowledge of mathematics, chance and motion. It also reinforces our view of a man hungry for data, with an eager appetite for the intellectual stimulation provided by original research and a tremendous enthusiasm for classification. The new science generated novel ways of looking at the world, and creating what we can think of as taxonomies of knowledge – of natural history, games or words – at the very heart of it. As many researchers will recognise, the mental exercise of classifying games almost certainly generated insights and ideas that proved useful in Willughby's classification of birds, fish and insects, and vice versa too.

Meanwhile, he and Ray continued to work through their ornithological notes, planning and preparing for the volume they anticipated producing. Had he known how little time he had left, Willughby would have undoubtedly been more focused on this particular enterprise.

8

Curious about Birds, Illness and Death

We have another brief flicker of pure light that helps to illuminate how Willughby's mind worked. This time it is in just two pages of the *Ornithology* that John Ray blandly entitled 'Some particulars which Mr Willughby propounded to himself to enquire out, observe and experiment in birds'. Ray included this list of questions posed by Willughby as he pulled everything together to produce the book. For me, these two pages are as exciting as the recent discovery that DNA can be extracted from the eggshells of New Zealand's extinct giant flightless moas.[1] I was surprised to find only one previous ornithologist who had noticed Willughby's questions and shared my enthusiastic opinion of them: William Jardine. In his 1843 memorial of Willughby, *Nectarinidae or Sun-Birds, with a Portrait and Memoir of Francis Willughby*, he wrote that these questions 'If founded on fact and drawn up with judgment, would not fail to contribute to the advancement of ornithology.' As well as doing this, they provide a rare opportunity to hear Willughby's voice.[2]

To be fair, the way Willughby phrases those questions isn't always straightforward – they were, in all likelihood, just jottings – and a superficial glance at them suggests that they might be cursory and of little consequence. But they really are, as Jardine said, an opportunity for enlightenment.

The puzzle is why Ray included them at all in the *Ornithology*. He introduces Willughby's questions in such an understated way and answers some of them with such weary indifference that I wonder why he bothered. But thank goodness he did. Lists of queries were beginning to be part of the way that scientists at the Royal Society tackled particular problems – as we saw with their botanical questions – and it is likely that Willughby created his list with exactly the same process in mind. It is evident also from his notes on sap in the commonplace book, from his *Book of Games* and from his legal papers, that asking himself questions was simply part of his modus operandi.

There are twenty-four ornithological questions in all, but there may well have been more, for several others are scattered through the body of the *Ornithology*. I have arranged them into several broad categories that include identification, reproduction, migration, age and the annual replacement of feathers.

Let's start with identification – the single most important goal of Willughby and Ray's enterprise and still the most significant challenge of being a birdwatcher today. If you cannot identify the species you've seen, you cannot tell anyone about it, and if you don't know what species you are studying, no one will take you seriously. Every birdwatcher and ornithologist will have experienced that sinking feeling when someone asks 'I've seen this unusual bird; it was brown and about the size of a pigeon ... can you tell me what it is?' Such imprecision is exactly what Willughby and Ray were desperate to avoid, and why, of course, Willughby took so much effort over his plumage descriptions. If, however, someone says 'I've seen this unusual bird almost as big as a cuckoo, with long wings like a martin, speckled like a woodcock ... with some white feathers on each wing' – much as Willughby's father-in-law had done in May 1668 – we would be fairly confident about identifying it as a European nightjar.

Today we use field guides to help us with bird identification. They work because they emphasise the key distinguishing features of live birds, either standing, swimming or flying, observed through binoculars. We could say that they identify those features that identify the bird. The first truly effective guide was Roger Tory Peterson's *A Field Guide to the Birds* published in 1934. Not only did Peterson excel as an artist, producing beautiful, clean, stylised images – in colour – that perfectly reduced each species to its most salient features, he also indicated by means of a short line the most important 'field marks' for identification.

Peterson's 'field marks' are similar to Willughby's 'characteristic marks' except that Peterson's consist only of those visible on a live bird. There were no binoculars in Willughby and Ray's day, and the telescopes that existed were so cumbersome they hardly lent themselves to observing birds.[3] The only way to really see and describe a bird (or any other animal) was to examine its corpse. The aphorism, 'what's shot is history, what's missed is mystery', sums up the pre-binocular era of ornithology.[4] With a specimen in the hand you had a reasonable chance of identifying it; if a bird simply flew past, you could never be sure what it was. Today's birdwatchers, conscious of the need for conservation, are justifiably critical of that earlier approach, but they often seem to forget that it was an essential part of the ontogeny of ornithology by which the study of birds came of age.

Of course, killing a bird destroys one of its most important characteristic marks, what's known as its 'jizz' – the shape, posture, mode of flight and so on. For most experienced birdwatchers today, a bird's jizz is as important for identification as its plumage. Interestingly, Francis Willughby recognised this for as we have seen, when constructing his classification of birds, he included behaviour as one of his criteria. John Ray, however, did not.

Imagine now that I have brought you, either from a seventeenth-century Italian bird market or from my freezer at home (where I've accumulated – quite legally – some birds found dead on the road), three birds: a willow warbler, a chiff chaff and a wood warbler. They are all very similar in size; they are all olive green on the back and they each have a pale stripe above the eye. With no previous knowledge, how confident would you be about saying whether these represented three different species or three individuals of a single slightly variable species? This was exactly the problem facing

all early ornithologists seeking to identify birds, and the only solution was to immerse yourself in the identification: to look at as
many examples as possible; to measure and scrutinise their external
features, noting the colour of the feathers, eyes and legs until such
time that your brain shifts into categorisation mode and the birds
suddenly start to form natural groupings.

I don't think we have any idea of just how difficult this was.
Perhaps only those twentieth-century birdwatchers lucky enough
to explore a remote part of the world with an unknown avifauna
can begin to imagine what it was like for Willughby and Ray. Even
then, our modern birdwatchers would have had the advantage of
being able to photograph a bird they caught to compare it with
those in field guides or museum specimens, and they may also have
recorded its calls and possibly taken a tiny blood sample to transcribe the bird's molecular signature.

What would Willughby and Ray have had to go on? They may
have carried with them on their travels one or two early bird books,
perhaps those by Conrad Gessner and Ulisse Aldrovandi. The illustrations in these inconveniently massive tomes were poor, and they
were in black and white. On the other hand, the authors' written
descriptions in both books were quite good and that must have
helped. Even so, time and time again in the *Ornithology*, Willughby
and Ray complain of the difficulty of matching up species they
have examined with those described by Gessner, Aldrovandi and
others whose work they quote. In addition, as we have seen, these
earlier authors weren't always especially careful and they sometimes
made mistakes with both descriptions and names.

This was why Willughby and Ray decided to start afresh, seeing
and describing every species for themselves, and crucially, doing so
in a careful, standardised way. Once viewed from this perspective
you can begin to appreciate why Willughby felt compelled to make
such detailed descriptions, and in the absence of reliable colour
images, to report colours in such precise ways. There were no colour standards in the 1600s (they didn't emerge until the 1920s), so
without an artist on hand to recreate the colours Willughby had no
option but to compare the colours he saw on birds with those of

familiar objects. Of the waxwing he says 'The outer wing feathers are marked with spots very pleasant to behold ... their appendices being red like to cinnabar or vermilion'; of the roller he says 'the wings are a lovely blue or ultramarine colour (as the painters call it) ... the head of a sordid green'; and of the dorsal surface of the song thrush, it is an 'olive colour, from its likeness to that of unripe pickled olives, such as are brought over to us out of Spain'.[5]

Over the years I have worked with several bird artists who have created images for field guides. Today's standards are high. You only have to compare the illustrations in the *Collins Bird Guide* (2009) with, say, the *Hamlyn Guide* that I used as a boy forty years earlier: the difference in the quality of both the painting and printing is remarkable. For the twentieth-century birdwatcher, those in the *Hamlyn Guide* are equivalent to Gessner's and Aldrovandi's impressionistic images of birds; those in the *Collins Bird Guide* are more akin to what Willughby aimed to achieve – a feather-by-feather accuracy.[6] Both Willughby and the Collins' artists, Killian Mullarney and Dan Zetterström, examined corpses – in Willughby's case they were freshly killed, in the later artists' case they were from museum drawers – to produce both a detailed written description and an accurate image. The main difference was that the Collins' artists had the benefit of large museum collections that allowed them to select and illustrate a typical specimen. They also benefited from photographs of live birds to visualise the perfect alignment of plumage, which isn't always the case on museum skins.[7]

⊹⊹⊹

One of Willughby's questions about characteristic marks was this: 'How many birds have an angular appendix, as it were a tooth, on each side of the upper chap of their bill, as the kestrel, the hobby, the butcher-bird, &c?'

He is referring here to a subtle feature known now as a tomial tooth, a hook on each side of the upper mandible, that occurs in two completely unrelated groups of birds: shrikes (butcherbirds) and certain birds of prey. Soon after I had written that sentence I found a dead red-backed shrike in France and was able to examine its beak and see its tomial tooth, which, I have to admit, I hadn't previously been aware of. The fact that this structure occurs in at least two unrelated groups of birds, both of which kill vertebrate prey, suggests that it is a killing device – an adaptation to a predatory lifestyle. Several recent studies mention the idea that raptors kill their prey with a neck bite that – facilitated by the tomial tooth – severs the spinal cord,[8] but the evidence is circumstantial. In shrikes there is a small amount of experimental evidence that this is indeed the function of this subtle notch: birds whose tomial teeth had been removed made less clean cuts and took longer to kill their prey.[9] Also, some other birds, such as owls and skuas that also kill vertebrate prey with a neck bite, have no tomial tooth, and a few other species also possess a tomial tooth whose function is unknown. Willughby's question almost certainly relates to the value of this feature in systematics and probably encouraged him to position the shrikes immediately after birds of prey in the *Ornithology*, although this is as likely to be because of their predatory behaviour as owing to the existence of the tomial tooth. Despite these similarities, shrikes and birds of prey are not closely related, and there is no hint nor indeed any expectation that Willughby or Ray recognised what we now call convergent evolution: the occurrence of similar traits in unrelated species with similar lifestyles.

The tomial tooth is a flagship feature for the issues of classification. From Willughby's day and before, people used anatomical features – both external and internal – as a way of classifying birds and other organisms. They had to: there was little else to go on. Willughby's idea that you could also use behaviour or song to classify birds was not taken seriously until the early 1900s when Oskar Heinroth and Konrad Lorenz recognised and exploited its potential.[10] Early naturalists assumed that God had imposed order on the natural world, and that their job was to figure out what God had in mind and reveal His arrangement. The most obvious clue was similarities in form: greenfinches and goldfinches are more similar to each other than they are to shearwaters or gulls. That much was obvious, and in theory at least it was seductively simple: one merely had to find those characteristic marks that both separated and united different species.

But it wasn't simple at all. As we now know, natural selection has created a multitude of forms, often riding roughshod over those features that might betray the true phylogenetic affinities of a species.

In the centuries that followed Willughby and Ray's pioneering work, ornithologists and other zoologists examined almost every conceivable external and internal feature they could think of to try to resolve the 'order' or classification of birds. They tried beak and bill structure, plumage pattern, eye colour, digestive tract layout (literally), the structure of the syrinx, as well as the presence or absence of rictal bristles (see below) and the occurrence of a tomial tooth, and many more features, but with only limited success. Revealing the 'arrangement' of birds dominated the subject of ornithology for more than 500 years and was the justification for vast Victorian museum collections.

Francis Willughby asks whether all individual birds of the same species have the same colour iris? In most cases, they do, with a small amount of variation, but in a few species the iris changes colour with age. In many gulls, for example, immature birds have a brown iris whereas in the adult it is yellow. Willughby also asks about what we now refer to as rictal bristles: 'How many [birds] have bristles under their chin, at the corners of their mouths, or about their nostrils?' The answer is quite a lot. They are especially prominent among nightjars, as recorded in the *Ornithology*, but rictal bristles aren't a defining feature of this, or any other, group of birds. Willughby also comments on the fact that many birds have a pale rump, and wonders too whether this might be a way of categorising them. But like all those other features he and subsequent authors identified, it isn't.

Interestingly, with all these questions, Willughby's focus is exclusively on identification rather than function. He doesn't ask *why* the shrike has a tomial tooth; *why* the irises of some birds – such as

those of giant petrels – are pale and seem so threatening, whereas
those of the giant petrel's close relative, the albatross, are dark and
seem so benevolent;[11] nor *why* nightjars have those bristles; or *why*
wheatears have a white rump. These are, of course, a different type
of question, and one that might not have entered Willughby's
mind. On the other hand, we know that John Ray was interested
in such 'functional' questions and spoke about them in his sermons and lectures in Cambridge, so it would be surprising if he
and Willughby had not at least discussed them.[12]

Let us turn now to Willughby's questions about reproduction. Two
of them concern the sex ratio of birds: one is about female reproductive anatomy and one asks whether the eggs of birds 'sometimes
fall from them against their wills'.

As has been known since ancient times, pigeons invariably lay
two eggs in a clutch, and Willughby asks whether it is true that
these give rise to one male and one female offspring. Aristotle and
the third-century AD Roman author Aelian both assumed this to be
the case, and rather surprisingly, so too did the normally cautious
William Harvey.[13] It is an intuitively appealing idea and one that
wouldn't be difficult to promote as an example of God's wisdom.
But it isn't true, as John Ray himself realised.

This is one of Willughby's questions to which Ray responds. He
writes that while you often do get one of each sex, sometimes the
eggs produce two males or two females. The truth is that, on average, 50 per cent of the time you get one of each sex, 25 per cent
of the time you get two males, and 25 per cent of the time two
females. This is the *overall* pattern, and it is what you would find if
you examined, say, 100 broods. It is a direct consequence of the two
sexes being equally common at the time of fertilisation.

In birds, unlike humans and other mammals, it is the female
that determines the sex of the offspring because the sex chromosomes reside in her ova (eggs). The sex chromosomes are assigned
to ova at random, which is why the sex of pigeon chicks hatching from the two eggs also occurs at random. When I first read

Willughby's query about this I desperately hoped that he might have understood why Ray's answer was correct. As we have already seen, Willughby was fascinated by games of chance, and it is chance that dictates whether you get one male and one female or two of each sex from a clutch of pigeon eggs. You can easily demonstrate this for yourself by placing an equal number of marbles (or indeed anything else), let's say 100 each, of two different colours in a bag; remove two at a time and note down what colour they each are. Replace the marbles each time and repeat this 100 times. You will find that you have close to fifty instances of one of each colour and twenty-five cases each when the two marbles are the same colour. This is known now as the 'binomial expansion'. However, being able to work out the sex ratio of pigeon chicks in this way, Willughby would need also to have known that in pigeons, and indeed in most sexually reproducing animals, equal numbers of males and females are produced at conception, but this wasn't generally recognised until the process of sex determination was established in the 1920s.[14]

That Francis Willughby was unaware of equal sex ratios among birds is clear from his next question. 'In what kinds of birds are there more cocks usually bred, as in ruffs, in what more hens, as in poultry?' The ruff is a wading bird that was once abundant in the marshy regions of Lincolnshire and may well have been among the species Willughby saw there in the summer of 1662. Ruffs were excellent eating and local Lincolnshire fowlers ran a cottage industry capturing and selling them. The birds were caught in clap nets, kept in the dark, and fed on a diet of bread and milk until they were fat enough for the table. Keeping them in the dark prevented them from seeing, fighting and killing each other. Willughby's Latin name for the ruff, *Avis pugnax*, says it all: the pugnacious bird. At least, the males are aggressive. The ruff has an unusual breeding system in which males congregate at a communal display arena (referred to as a lek) where they compete aggressively for a tiny patch of ground that provides them with the opportunity to copulate with females on their visits to the lek. After copulating, the female – known as the reeve – goes off to incubate her eggs and rear her chicks alone. Most of the time, therefore, more males than females are present on the lek, and hence when the Lincolnshire trappers caught ruffs at their leks they generally secured more males than females, giving the impression that the sex ratio was strongly biased towards males. The intense competition between males for matings on the lek has resulted in their evolving both aggressive behaviours and also elaborate head and neck plumes – the ruff. Because it was once believed that wild populations of ruffs contain more males than females, there have been several attempts to assess the 'real' sex ratio of ruffs. Contrary to expectation, these all indicate a slight preponderance of females in the adult population.[15]

Willughby's idea that in poultry – by which we can presume he means domestic fowl – females outnumber males is also a consequence of their mating system. Typically in domestic fowl, and in their wild ancestor the red jungle fowl, a single dominant cockerel assumes control over a harem of several females, driving away the younger, more subordinate males. But the sex ratio among adult fowl is close to even. Indeed, if the modern poultry industry had found a way to break the biological fifty-fifty stranglehold of this aspect of avian reproduction, they surely would have done so because the profit is largely in hens rather than cockerels.

This fascination for sex ratios may have had its origin in Venice when Willughby and his colleagues were there, discussing the slight excess of human males revealed by the 1581 census. Ray comments that this is similar to what has been found in London: 'And I doubt not but if exact observations were made in other places, there would be found the like [same] proportion between the numbers of males and females born in the world in hot countries, as in cold, so that from this topic, the Asiaticks [sic] have no greater plea for multiplicity of wives, than the Europeans.'[16]

ᛡᛏᛡ

Willughby and Ray made a point of dissecting as many bird (and fish) species as possible, noting the state of their internal organs and the gonads, in particular. They were quick to comment on whether a male's testicles (testes) were relatively large (as in the quail, house sparrow, dunnock and turtle dove) or small (as in the shrike they killed near Augsburg). Exceptionally large testes tell us something interesting; small testes don't necessarily indicate much because in all birds that breed in temperate regions, such as Europe, the testes shrink away to almost nothing outside the breeding season and this means that a bird with tiny testes may simply be out of breeding condition. A bird with big testes, however, is one in which the females of that species are promiscuous.

Allow me to explain the logic here. As a result of molecular paternity studies conducted during the 1980s and 1990s, it is now well known that, despite the fact that most birds (but not ruffs or jungle fowl) breed as pairs, infidelity is common and results in broods of mixed paternity. In species where females are unfaithful, males attempt to minimise their loss of paternity by transferring more sperm, and they do this by evolving larger testes. Willughby and Ray naturally knew nothing of evolution, nor the concept of what is known as 'sperm competition', but they were on the right track when they said that the quail had 'Great testicles for the bigness of its body, whence we may infer that it is a salacious bird' – as indeed it is. Frequent copulation with his partner is the male quail's best chance of being the biological father of her offspring.[17]

Like the testes, the reproductive organs of female birds – the ovary and the oviduct – also show profound changes in size through the year, and are at their maximum, of course, during the breeding season. The autumnal reduction in size may be an adaptation for flight – why carry all that additional anatomical baggage outside the breeding season when it isn't needed? A further weight-saving adaptation in female birds is achieved by having – in most species – only a single functional ovary and oviduct. In comparison, most mammals and reptiles have two – a paired system. The existence of just one, and it is usually the left ovary and oviduct, has been known for as long as people have been butchering chickens. But as Willughby's question suggests, this might not always be true. He asks: 'Whether some birds have a double cluster of eggs, as viviparous animals have two ovaria?'

Double ovaries have been recorded in two groups of birds: in raptors, such as the Eurasian sparrowhawk and harriers, and in kiwis – routinely.[18] It is possible that Willughby found double ovaries in some of the raptors he and Ray dissected. Kiwis weren't discovered until the nineteenth century, so he couldn't have known about them. On reading through the raptor accounts in the *Ornithology*, I was disappointed to find not a single instance where Willughby had discovered a double ovary. It is possible that one of the earlier ornithologists, such as Gessner, Belon or Aldrovandi, had noted an example, but I had no luck there either. It is not known why some female raptors occasionally have a double reproductive system, but in kiwis (which of course are flightless and so do not benefit in the same way as flying birds from saving weight) it is thought that the two ovaries and oviducts alternate in producing eggs. But of course this still does not explain their existence.[19]

Willughby's final question about reproduction asks whether female birds that are ready to lay eggs are able to hold on to them if, for example, their nest is destroyed, or 'whether they [their eggs] sometimes fall from them against their wills'.

The only thing I can think of that might have prompted this question was that Willughby, or someone he knew, had encountered isolated eggs in unlikely locations – such as an intact starling egg lying conspicuously in the middle of a lawn. As a boy, when starlings were much more common than they are today, I came across several such eggs, and like Francis Willughby I wondered whether the female had been 'caught short' and forced to lay away from her nest. I couldn't think of any other explanation. However, in the 1970s, careful field studies showed that starlings are unusual in sometimes behaving like a cuckoo, laying eggs in the nests of other females, but of their own species. Some female starlings attempt to exploit the parental care of other starlings by dumping one or more eggs in their nest. As they do so, the parasitic individual removes one of the host's eggs (presumably to minimise the chances of her egg being detected) and then flies off with it and drops it somewhere.[20] Eggs dropped onto a hard surface smash and quickly disappear; those dropped onto a grassy lawn are more likely to survive intact.

Another example of eggs turning up in unlikely places is guillemot eggs dredged up from the seabed by fishing boats many miles from any breeding colony. Such eggs were greatly prized by nineteenth-century egg collectors who, of course, were desperately vulnerable to being duped. However, there were enough records to show that guillemots must occasionally – for whatever reason – lay their eggs far out at sea, exactly as if they have been 'caught short'.[21]

Experiments on chickens have shown that if a hen is deprived of somewhere to lay, she can hold on to her egg for several hours beyond her normal laying time.[22] So, this answers Willughby's question: birds can hold on to an egg for a while, but after a certain time, if they are unable to lay in their nest, they are forced to deposit their egg more or less at random.

There is one bird, however, that routinely holds on to its egg for longer than most others, the European cuckoo. As Willughby and Ray knew, this species is a brood parasite, depositing its eggs in the nests of other species so that the young cuckoo is reared by foster parents such as the dunnock. Whether Willughby and Ray knew that Aristotle had commented on the cuckoo's parasitic habit isn't clear, but they were, nevertheless, shocked by it because, as one of them – probably John Ray – writes, it 'seems so strange, monstrous and absurd, that for my part I cannot sufficiently wonder there should be such an example in nature; nor could I have been induced to believe such a thing had been done by nature's instinct, had I not with my own eyes seen it.'

One key to the cuckoo chick's success is hatching before those of its host. This allows the young cuckoo to eject the host's eggs or chicks, and to monopolise the food brought by the parents. Early hatching is arranged by the cuckoo in several ways: by laying early in the host's laying sequence; by producing a relatively small egg that requires fewer days incubation; and partly by giving her egg a head start by retaining and incubating it within her body for an additional twenty-four hours. Laying early in the host's sequence and producing a small egg are both adaptations that have evolved through natural selection as part of the cuckoo's brood parasitic repertoire. Retaining the egg in the oviduct for an extra twenty-four hours is a fortuitous accident (for non-parasitic species of cuckoo also do it), which may nevertheless have predisposed cuckoos to become brood parasites.

In most birds, including the domestic fowl and songbirds such as thrushes, wagtails and pipits, females lay their eggs at intervals of twenty-four hours. That's how long it takes for an ovum to be released from the ovary, be fertilised, to have the albumen and shell added to form the egg and to pass through the oviduct and into the nest. The ovum (essentially what we think of as the yolk) is fertilised within minutes of being released from the ovary and the embryo's development starts a little bit further down the oviduct about five hours later. By the time the completed egg is laid, the embryo consists of a few thousand cells that appear as no more than a pale speck on the yolk's surface. Things are different in the cuckoo; its eggs are laid at intervals of forty-eight hours. Each egg takes twenty-four hours to form, just as in other birds, but the female cuckoo retains the fully formed, shelled egg inside her for a further twenty-four hours. During this time the embryo contin-ues to develop, so that when the egg is laid the embryo has a head start. An additional twenty-four hours in the oviduct might be expected to provide the chick with a twenty-four-hour advantage, but because the body temperature (40°C) of the female cuckoo is several degrees warmer than eggs experience during incubation (36°C), the young cuckoo ends up with a thirty-hour hatching advantage over the host chicks.[23]

Willughby's final three questions are about how birds survive the cold of winter conditions; which of them migrate or hide; and what would become of migrant birds if they were kept in captivity over the winter?

For small birds that don't migrate, the way they cope with cold winter nights is truly remarkable, and I think Willughby would have been astonished at what ornithologists have since discovered. This is how Willughby phrased his question: 'How cometh it to pass that the most vehement cold in winter-time, if they have but food enough, doth not congeal or mortifie the tender body of small birds?'

Before we answer this, let us consider what our smallest European bird, the goldcrest, has to endure. Weighing just five grams – little more than a teaspoon of sugar – a goldcrest typically has to survive sixteen hours of winter darkness each night at temperatures as low as minus 20°C. How does it do that?

Superficially, we might imagine the answer to be: fat and feathers. Even Willughby might have realised this, but in fact the way small birds make it through the long winter nights depends on an extraordinary combination of anatomical, physiological and behavioural tactics.

Birds have to possess sufficient fuel in the form of fat to keep their engines generating heat overnight. Many small birds forage strategically during the winter day, subconsciously deciding on the basis of the ambient temperature how much food to convert to fat for their overnight survival. It is a delicate balance: too little fat and you don't make it through the night; too much and you are too heavy when flying and vulnerable to being caught by a hawk.

Feathers provide extraordinary insulation, and birds typically fluff themselves up into a ball when they go to roost. This is most easily seen if you have a camera nest box used by a roosting great tit or blue tit during the winter. That spherical shape minimises the bird's surface area, and hence its heat loss. The colder it is the more the feathers are erected and the more air is trapped and the less heat lost. It is often said that birds put their head under their wing when they sleep, but that's not true: rather, they turn their neck and place their head under the feathers on the back. With the rest of the plumage fluffed up, the head becomes invisible, and there's a good reason for this. More heat is lost from the head, and especially around the eyes and beak, than from anywhere else on the bird's body, so keeping the head warm is a priority.

Roosting in a nest box isn't a bad idea, for the microclimate at the roost site obviously has a huge effect on the temperature to which a sleeping bird is exposed. Many small birds roost in cavities and some – such as tree creepers and wrens – do so in huddling groups, benefiting from each other's body heat and insulation. Sometimes goldcrests roost in the open, but usually with others in a huddle, for the same reason.

We know from our own experience of trying to stay warm in bed that food and feathers (an insulating eiderdown or down-filled duvet) are essential. Willughby would have understood this, but he would also have recognised that our problem isn't as acute as that of a goldcrest because our greater body size means that we lose heat more slowly. There must be something else going on, and indeed there is, but I doubt that Francis Willughby could have anticipated what it is.

Studies of the North American black-capped chickadee by Susan Budd Chaplin in the 1970s revealed that although they went to roost with some 7 per cent of their body mass as fat, this was insufficient to maintain their body temperature (42°C) and survival through the entire night, despite their wonderful insulating feathers.

What Chaplin discovered was extraordinary: the chickadees drop their night-time body temperature to around 30°C, entering a state of hypothermia and reducing their fuel consumption so that there *is* sufficient fat to see them through the night. Hummingbirds, which are even smaller, and are especially vulnerable to losing heat at night (even in the tropics, and especially at high altitudes), do something similar and drop their body temperature to below 10°C, becoming torpid as a result. The energetic saving of torpor is enormous. Torpor also occurs in both the European nightjar and the common poorwill of North America, which in the latter case can remain in a state of suspended animation for several days on end – a mini-hibernation. Even birds such as tits and finches reduce their roosting body temperature – usually by only five degrees – to enjoy the benefits of this overnight energy saving.[24]

⋔

What about migrants? In Willughby's day the whole issue of bird migration was still up in the air, so to speak. Doubt about migration had been cast centuries before by Aristotle, who suggested that some birds metamorphosed into others between summer and winter – thereby accounting for why some species disappear and others appear in winter. But he also thought that some hibernated, hiding themselves away in holes during the winter months. He was fairly confident that others, such as white storks, migrated to warmer climes, because he could see them leaving in the autumn and returning again in the spring. Willughby agreed, saying in the *Ornithology*: 'it is most certain, that storks before the approach of winter fly out of Germany into more temperate and hot countries'. There were a few other convincing instances of migration, including the ornithologist Pierre Belon's observations of common quail made in the mid-1500s: 'When we sailed from Rhodes to Alexandria of Egypt many quails flying from the north towards the south were taken into our ship, whence I am verily persuaded that they shift places.'[25] The real problem was small birds such as warblers and swallows. It was obvious that they disappeared in the autumn, but they seemed too fragile to fly across vast tracts of ocean, and – because they typically migrate at night – their seasonal movements remained unobserved. It was precisely for this reason that doubts about migration in such species persisted for so long. It wasn't until the 1800s that most naturalists eventually agreed that migration was the only plausible explanation for their seasonal appearance and disappearance.

The answer, therefore, to Willughby's question – 'What birds hide themselves or change places, whether in summer or winter?' – is that no birds hide themselves or hibernate; those like warblers, nightingales and flycatchers migrate, while others such as finches and tits stick it out and are resident all year round.

The next part of Willughby's question is a revealing one: 'What would become of nightingales, cuckoos etc., in winter, and old fieldfares &c. in summer if they were kept in cages, and carefully tended, fed and cherished?' I should first point out that – as Willughby was well aware – the fieldfare is a winter visitor from northern Europe to Britain, which is why he asks what would happen to it in summer. Willughby's question is revealing because it confirms a suspicion I have from reading the *Ornithology*, that neither he nor Ray had very much to do with those people who trapped or kept small birds as cage birds. Keeping songbirds – including canaries imported from Spain – was increasingly popular from the Middle Ages onwards, and a huge variety of species was kept. Most of them probably didn't survive very well, but some – including a few migratory species – occasionally found themselves in the care of an expert and lived long and lusty lives. This means of course that had Willughby been more familiar with this aspect of ornithology he might have been able to answer his own question.

Of all the small birds kept as pets in the seventeenth century, the nightingale – a migratory species – was the most popular, because of its lovely song. Also, as Willughby and Ray knew, Valli da Todi and Olina had already written extensively about the habits of this species. And although they do not say so explicitly, my guess is that both these authors – and probably other individuals too – will have kept nightingales over winter.[26] Cuckoos would have been a different matter, because everyone that has tried it has found them to be impossible to maintain in captivity.

The first details of what captive nightingales do during the winter came from Nicolas Venette, a French physician and professor of anatomy at La Rochelle, whose account appeared in 1697 while John Ray was still alive. Venette was a polymath and published on a wide range of topics including pruning trees, poetry and the treatment of scurvy. He was best known, however, for a work entitled *De la génération de l'homme, ou tableau de l'amour conjugal*, popularly known as *Conjugal Love*, a book that dealt with genital anatomy, sexual gratification, pregnancy and embryo development. Most references to it criticised the book for its lewdness, but the English translation I obtained revealed it to be anything but lewd. Instead *Conjugal Love* is a wide-ranging sex manual that includes genuine attempts to solve problems associated with pregnancy and sexual dysfunction – something with which Ray, given that he had his own period of impotence after his marriage in 1673, should have been able to identify.

When Venette retired from medical practice, he was evidently not very mobile and decided to study captive nightingales. Among the many things he noticed was their agitated behaviour in autumn and spring at exactly the time when wild nightingales were migrating. Perceptively, he correctly deduced that the birds' frantic hopping was sublimated migration and, indeed, this behaviour later came to be known as 'migratory restlessness'. However, because it was assumed to have been discovered by German bird-keepers in the 1700s, it was termed 'Zugunruhe', but in fact our French physician seems to have been the first to document it towards the end of the preceding century. Subsequently, many of those writing about cage birds, such as quail and the golden oriole, commented on their agitated behaviour at the time of migration. Venette himself was convinced that small birds migrated and that an internal clock controlled their seasonal restlessness. He also made some ingenious suggestions regarding the cues that birds use to find their way on migration.[27]

I'm convinced that Willughby and Ray missed a trick by not paying more attention to bird-keeping. The English edition of the *Ornithology*, as we've seen, did include a considerable amount of bird-keeping information – mainly from Aldrovandi's *Ornitholigae*, Olina's *Uccelliera* and Gervase Markham's *Hunger's Prevention* – added at Martin Lister's suggestion, but only after the main research and writing was complete. This means that even though the *Ornithology* includes references to some intriguing aspects of bird biology derived from cage birds, including the acquisition and function of song, this information is incidental and is not fully integrated into the book's main themes of identification and classification.[28]

You can tell from their correspondence that Ray was reluctant to include this extra material, but Lister was convinced it would increase the book's popularity, and Willughby's widow, it seems, was hoping that the book might yet yield a profit. Lister clearly recognised bird-keeping as an important aspect of popular culture – why else suggest including a section on this? – and he was sufficiently perceptive to realise that it could reveal some new insights. But Ray either failed to see the point, or was in a hurry to complete the English edition.

ↆↀↆ

Hidden away in the pages of the *Ornithology* are other questions posed by Willughby. Here are just two examples of queries that do not appear on Ray's list.

The first relates to moult: the annual replacement of feathers. Ray writes in the *Ornithology*:

> It may be worth the while to enquire, why birds do yearly moult their feathers? Mr Willughby supposes that there is the same cause of the casting of feathers in birds, that there is of the falling off of the hair in men and other animals...[29]

So far so good. Two facts, both correct, are established: that the moulting of feathers and hair is an annual event, and that it is the same in birds and mammals. But then, as Ray tells us, Willughby goes on to say:

> ... upon recovery from fever or other disease, or upon resection after long abstinence. For in cock-birds the heat and turgency of lust is, as it were, a kind of fever, and so in the spring-time their bodies being exhausted by the frequent use of venery, they become lean: but in the hens the time of sitting and bringing up their young answers to a disease or long abstinence, for at that time they macerate themselves by hunger and continual labour. When these times are over both sexes returning to mind their own bodies and feed for themselves, do in short time recover their flesh and grow fat again, whereupon the pores of the skin being dilated the feathers fall off.[30]

Willughby's idea – and it is a rather strange one – is that the loss of feathers (and hair in humans) occurs when an individual recovers its health following a period of stress, which in turn could be a consequence of disease; too much sex in the case of males; or among female birds the energetic cost of incubation, brooding and chick-rearing. What Willughby means by 'upon resection after long abstinence' could either be the resumption of eating after fasting, or the resumption of sex after abstinence. His statement seems to be a badly remembered or misconstrued account – and certainly not a quote – of something Aristotle said. If it is, I have been unable to find it. Instead, what I discovered in Aristotle's books on *Generation*

and the *History of Animals* was a presumed link between the growth
and loss of hair with lust, but also the similarity between the falling
off of hair and the falling of autumn leaves in Aristotle's desperate
attempt to understand the appearance, disappearance and greying
of human hair.[31]

Willughby is correct, however, in stating that moulting follows
reproduction – although this was already well known by numerous
medieval writers. Such a link is particularly obvious in male ducks
when they exchange their bright breeding plumage for more cryptic garb, as they, and the females, shed their flight feathers, thereby
becoming flightless for a few weeks in the autumn until their new
feathers have regrown.

Willughby was not the first to discuss moult. Frederick II, in his
thirteenth-century account of falconry, has quite a lot to say about
it, and is actually much more perceptive than Willughby. But it
is extremely unlikely that Francis could ever have seen Frederick's
account because it was not published until long after Willughby
and Ray had died.[32] Nonetheless, other falconry treatises that
would have been available to Willughby (and Ray) mention moult,
not least because it was such an important aspect of maintaining
raptors in good condition. But it seems that neither Willughby nor
Ray originally consulted any of these, and Ray did so only after
Francis was dead when he added an account of falconry to the
English translation of the *Ornithology*. Even then, Ray writes very
little about the loss and replacement of feathers.

The absence is curious because moulting, or mewing, as it was
known, was a critical period in the falcon's annual cycle. Falconers
were very familiar with the pattern of feather loss and replacement
in their birds: in fact, the terms 'moult' and 'mew' both come from
the Latin *muto*, meaning 'to change', hence the mews as a place
where hawks were housed while they moulted. One of the books
that Lister encouraged Ray to cite in the English edition of the
Ornithology was *Latham's Falconry or the Falcons Lure or Cure*, published in 1615. Latham's account includes one fascinating piece of
advice that I'm surprised Ray did not comment on. It was the falconers' practice of feeding their moulting birds on 'kirnels', the

thyroid glands of sheep that were thought to help birds through the moult. This sounded like another old wives' tale to me, one that Thomas Browne might have considered a 'vulgar error' and been keen to debunk, but in fact it works. The sheep's thyroid gland contains the hormone thyroxin that in sheep controls their metabolism, but in birds stimulates moult and feather growth. Latham of course had no notion of hormones, but this was clearly an effective bit of folklore.[33]

I suspect that because moulting is a *process* lasting weeks, or sometimes months, it was a subject that Willughby and Ray were unable (or unwilling) to investigate because it would have entailed keeping birds. Had they had cage birds, or even falcons, they would have had a better grasp of the process. Perhaps also if Willughby had appreciated that the *pattern* of moulting – the order in which wing feathers, for example, are shed and regrown – differs so much between different groups of birds, and if he had recognised that these patterns *may* have been a potential clue to the classification of birds, he might have taken more of an interest. Certainly, this is exactly what he did with insect metamorphosis, but then keeping insects was rather easier than keeping birds. It was the idea that moulting patterns might reveal hitherto unidentified taxonomic relations that attracted Erwin Stresemann and his wife Vesta to this area of ornithology in the 1950s. But, as must quickly have become apparent, moult did not inform classification, and their 1966 book *Die Mauser der Vogel* (*The Moult of Birds*) includes only a few pages on taxonomy.[34]

Willughby's list of ornithological queries must have once been longer than that which appears in the *Ornithology*. In the section on the common (or grey) heron, Ray states that 'Gessner counts but eleven vertebres [sic] in the neck; I observed fifteen, of which the fifth has a contrary position, viz, is reflected upwards.' Ray adds a note in the margin, saying 'In another place Mr Willughby puts it among his queries, whether the five upper vertebres [sic] in the neck of a heron be reflected the contrary way.' This sounds like a gentle admonishment that asks whether Willughby means just the fifth vertebra, or the upper five vertebrae that are reflected upwards? It may be because of that ambiguity that Ray did not include this particular query in the main list of queries. Given that several of Willughby's other queries elicited responses from Ray, it is interesting that he doesn't tell us the true answer, which makes me wonder whether he knew, and therefore whether anyone else knew at the time. Indeed, I think that Willughby may have been the first to describe these unusual vertebrae in the necks of herons. It is also intriguing that neither Willughby nor Ray say anything about the inflected vertebra being responsible for the distinctive kink in the heron's neck, both of which seem such obvious questions, albeit in hindsight.

The numbering of cervical vertebrae in the grey heron by Willughby is actually correct, except that like some other authors, he did not count the first rather small vertebra, the atlas – the one immediately adjacent to the skull. He therefore identifies the special vertebra as the fifth one, whereas if you count the atlas as number one, it is the sixth. We also know that by being longer and articulating with the next vertebra at a different angle, it is this

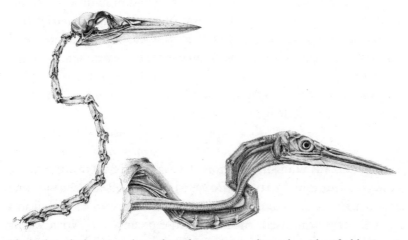

The kink in the heron's neck, resulting from a 'reversed' vertebra, identified by Francis Willughby.

sixth vertebra that is responsible for the neck's distinctive kink. And that, acting rather like a spear-thrower, allows the heron (and several of its close relatives, including egrets, bitterns and the darter) to more effectively stab its prey.[35] I am intrigued to know how Francis Willughby made this vertebral discovery, for in a routine dissection, that is, without removing the muscle, the orientation of the neck vertebrae is not obvious. Rather, I suspect that he noticed the unusual bones while examining some of the mounted skeletons he and his colleagues encountered during their continental travels.[36]

Overall, those two pages of Willughby's questions allow us to detect someone with an eager, enquiring mind; someone pushing the boundaries and a little careless perhaps of staying within the limits of his and Ray's original aims. As today's researchers know all too well, being curious about the natural world and constantly asking questions is an essential aspect of being a biologist. It is what Darwin identified in his brief autobiography as one of the attributes that made him what he was. I still find it puzzling that John Ray was so obviously vexed by Willughby veering into unknown territory, and I can only think that it was because, in his own inimitable way, Ray had a very clear vision of what the *Ornithology* had to do, and was worried that his colleague was trying to fly before either of them could walk. Those questions, however, suggest to me that had Willughby lived longer he would have gone on to think and write about birds in an extraordinarily novel way, with much more emphasis on their lives than their bodies.

Throughout his adult life Francis Willughby seems to have had a weak constitution. At Trinity College his tutor, James Duport, was acutely aware of this and penned poems urging him to slow down and take care of himself. Although those poems were written after Francis had left Cambridge, they were clearly based on his time there and I can only presume that Duport also told Francis to his face that he should work less hard. But of course, if that's the way one is, it is hard to stop doing what one loves.

Willughby's first bout of serious illness that we know of occurred in June 1662 when at the age of twenty-seven he was returning from Wales with John Ray and Philip Skippon. At Malvern, Francis became unwell and needed two days of bed rest and fasting before the 'heat of the distemper' subsided sufficiently to allow him to return home.

We next hear of Francis being ill on several occasions between 1668 and 1671, in his daughter Cassandra's account of his life written long after her father's death. She relates, presumably based on family diaries, how he suffered violent fevers, and sought relief by taking the waters both at Astrop and at Cleehills near Ludlow while staying with the politician Sir Job Charlton.

Francis was taken seriously ill again in 1669 while he and Ray were visiting John Wilkins in Chester. We know little about this particular bout of illness other than its debilitating nature. The following year there was another bout and on 28 April 1670, Ray wrote to Martin Lister saying that 'Mr Willughby is I fear, fallen into a tertian ague.' It may have been Willughby's recovery from this malady, or possibly another one, that Ray told Lister about, in a letter dated December 1670, to which Lister replied saying: 'I am very glad Mr Willughby is neer well again, and I thank God for his recovery, and doe heartily pray a continuance of good health to him. Methinks he is very valetudinary, and you have often alarmed me with his illness ...'[37]

A year or so of reasonable health followed his recovery, but all was not well. On 10 February 1671 a distant relative, Sir William Willoughby of Selston, died, leaving Francis as part-heir to a considerable estate at South Muskham and Carlton in Nottinghamshire, lands worth £1,000 a year (equivalent to around £80,000 today). It wasn't, however, straightforward. Part of the family estate, by earlier legal arrangements, descended to Sir William's sister and her husband, Beaumont Dixie. They had expected to inherit the bulk of Willoughby's lands and money, and argued that Sir William had not been of sound mind when he made the will. Worse still, and revealing the man he was, Dixie went into the house before his brother-in-law had even died, removing papers and goods and helping himself to several thousand pounds in cash. As soon as Sir William died, Dixie placed padlocks on the doors to prevent the executors from making their inventory.

Francis was summoned, and went to Selston with John Ray where they met the executors and sought legal advice. In a massive understatement, Ray described Dixie as being 'of a contentious spirit', and in a letter to Martin Lister he described how Francis was so preoccupied by this conflict with Dixie that he had little time for his studies: 'The estate bequeathed him, will create him a great deal of trouble, and cost much money, and yet what the issue may be is uncertain, by reason of an errour in the will.'[38]

Ray was correct: the situation was complicated, but one whose long-term advantages were considerable for Francis's family, so it isn't surprising that he committed a huge amount of time and emotional energy to resolving the case. At one point in 1671, knowing of Dixie's 'designs to get the will from him', Francis was forced to ride the 140 miles to York at high speed to 'get the will proved'. Later that same year he had to go to York again, to collect the will and take it to London. Travelling by coach via Nottingham and Haslington in Cambridgeshire, Francis picked up his brother-in-law, Sir Thomas Wendy, on the way, to help him defend his case. It was physically exhausting and financially crippling. When the accounts were drawn up later, it became clear that Willughby had spent £1,509 9s 2d – over £100,000 in today's money – on the case. Yet Dixie didn't give up, the case dragged on, and it was not until the eighteenth century that the inheritance was eventually secured.[39]

Francis and Emma's second son, Thomas, was born in April 1672. This was an occasion for celebration and a wonderful distraction – albeit briefly – from the stress of contesting the will, running the estate, and failing to be able to get on with his studies.

In early May, Willughby was in London at the Royal Society, planning another botanical tour to the West Country, and able to comment on a letter sent in by Martin Lister on parasitic worms. Francis told the Society how he had regularly encountered such worms in many of the animals he had dissected – including a cormorant chick in the Netherlands that contained a selection of black worms – adding that he had found them both in the gut and also 'lying loose in the cavity of the belly'. Parasitic worms, we now know, most often inhabit the gut, but when the fish, bird or quadruped host dies, the worms attempt to escape, burrowing through the gut and into the abdominal cavity. Francis promised to send the Society the names of all those animals in which he had found worms, but illness intervened, the plant-hunting trip was abandoned, and the list of parasite hosts was never sent.[40]

On the third day of June 1672, Francis awoke next to his beloved wife, knowing that he was sick once more. Dr Anthony Hewitt, a Padua-trained physician, called from Lichfield, would have taken Francis's pulse, examined his urine, removed several ounces of blood, and probably prescribed a purge and an emetic – all in the hope of restoring the imbalance between Francis's humours. Seventeenth-century medicine knew no better and had precious little to offer in the way of either relief or reassurance.

Two weeks after the first attack, Francis acknowledged to John Ray that he was not going to recover. They discussed what was to become of their joint endeavours, with Ray reassuring Francis that he would pull everything together and ensure that all would be published. Francis told him that the insect work – the caterpillars, the little butterfly maggots, the beautiful little bees in the willow wood – were what, above all else, he wanted in print. He also asked Ray to look after his sons' education, and Ray agreed.

On 24 June Francis signed his will, which among other things declared him a Protestant 'utterly detesting and abhorring all the idolatory, abominable errours and ignorance that is the Popish religion', and barring inheritance by any descendant who was a Catholic. Francis made provision for his family, optimistic that the Dixie affair would be resolved in their favour, but prudently making provision in case it wasn't.

He must have known the end was near. Every day the family – the children now dressed in black, anticipating the inevitable – and

John Ray came to his room, and each day they must have later walked away with a sense of hopelessness. The maids sat in attendance day and night. Emma's drawn and tear-stained face must have upset Francis, but as though to reassure his wife, he admitted to her and Ray that if it pleased God he was content to leave this world.

And so he did, on the morning of 3 July 1672, thirty days after first becoming ill. He was just thirty-six.

Many years later Emma recounted the events of those dreadful days to her daughter Cassandra, who recorded them thus:

> He was seized ... with a plurisie, which terminated in that sort of fever which phisitians [physicians] call catarhalis, which illness he labored under a month. He very soon apprehended his own danger ... expressed great concern about leaving his children and parting from my mother [Emma, his wife] who he had lived with about four years and a half – declaring that during this entire time no unkind word had passed between them.[41]

If pleurisy is what killed him, it comprised an inflammation of the membrane around the lungs causing chest pains and breathing difficulties: symptoms of what in all probability was pneumonia. Francis may also have been suffering from what was known then as 'tertian fever', ague or malaria, contracted either in the marshes of Lincolnshire or during his travels in Italy, and accounting for the alternating bouts of shivering and fever. Add to this a constitutional fragility and all the physical and mental stresses associated with the inheritance, and it is little wonder his body gave up.

Francis's funeral took place at Middleton parish church a few days after his death, with the family's curate and friend George Antrobus preaching the final sermon. At the graveside with the family were Francis's dear friends, John Ray, Philip Skippon and Francis Jessop.

9

Into the Light: Publication

Francis Willughby's death caused not only despair among his family, friends and Royal Society colleagues, it unleashed decades of difficulties for them all. Following the funeral, Willughby's father-in-law, Henry Barnard, took control. On discovering that there was a mere £160 of ready cash left in the coffers, his immediate task was to borrow monies – including some from Francis's mother, Lady Cassandra – to keep the household afloat. John Ray, together with Francis's friends, Philip Skippon and Francis Jessop, and Willughby's brother-in-law, Thomas Wendy, were executors, but Barnard felt that none of them – and Ray in particular – fulfilled their duties.

For his part, Barnard assumed that the £60 Willughby had granted Ray as an annuity meant Ray would remain as part of the household, educating Willughby's sons Francis and Thomas, and – almost incidentally as far as he was concerned – bringing Francis's notes to a form in which they could be published. By the time Francis died, he and Ray had already spent seven years working on their ornithological materials, but they were by no means ready for publication. Ray's annuity was fairly modest; sufficient perhaps if he was to be housed and fed in the Willughby home, but quite inadequate if he was to live independently, when a sum of around £200 per annum would have been more reasonable.

In Barnard's eyes, Ray needed to earn his annuity by taking an active role in running the estate, especially since the family knew

that he and Francis had spent a considerable amount of time studying the legal records of the Willoughbys' property and lands. Even
less realistically, Barnard felt that Ray should also help to pursue
the ongoing inheritance issue with Beaumont Dixie.

Initially at least, Ray had the support of Lady Cassandra, enabling
him to continue to focus on what he considered his main responsibility – preparing the *Ornithology*. His relationship with Emma,
however, was more problematic. Although she agreed to cover the
cost of engraving the plates for the *Ornithology*, she did so on the
assumption that its publication would yield a profit – which was not
unreasonable, given the dire state of the family's finances. Her feelings towards Ray, Jessop and Skippon, however, were soured by the
sense that between them they had selfishly consumed valuable time
she could have spent with her husband, and that by encouraging his
studies they had somehow contributed to his death.[1]

⊦⋎⊦

For months, Emma was unable to control her sobbing. Francis's
death left her feeling cheated: four and a half short years of marriage, with so much to look forward to, now lost. The family was
worried by her state of mind. Six weeks after the funeral, her sister-
in-law Lettice, anxious that Emma was giving way to 'excessive
grief', wrote to her saying: 'Take heede you provoke not God to
goe further on in smiteing; and though you have lost ye greatest
losse yet doe not think you have lost all.'

Half a year later, Emma's father wrote to her concerned by her
'melancholy humour' and the fact that she had 'little mind of doe-
ing anything but what necessity compels you'. Reminding her that
she had three young children to care for, he urged her to cheer up
her spirits. But Emma's spirits took another knock when her other
sister-in-law, Francis's sister Katherine, then living at Middleton,
lost her own husband.[2] Emma's grief was protracted and a third of
a century after Francis's death her memory of him remained vivid
and her loss acute.

Still anxious over Emma's state of mind, Lettice wrote to her
in December 1673 offering to bring her a tame magpie that she

thought would cheer her and the children. A friend had a talking magpie he wanted to get rid of, 'kept in a cage and twill speak many words very plaine'. Lettice thought the children would 'delight in it', adding 'as I did myself when the days were that I delighted in anything'. She then says that she would have brought it without asking, but was then concerned that Emma might be worried that the bird would distract the children from their lessons, 'besides tis a little curst [cursed] if [one] come[s] too neer his cage [he] will put out his bill and catch at fingers or anything he can lite on, but that is easily prevented with care. Let me have an answer for I stop disposing of him till I hear.' Magpies do have a surprisingly hard nip, as anyone who has caught a wild one, or had one as a pet, will testify. Fewer will have heard a talking magpie, and Lettice is right, it would have delighted Emma's children, for the voice in which magpies speak is utterly bewitching – a friend of mine had one that recited several nursery rhymes in their entirety.

It isn't known whether the magpie made it to Middleton. It may have done, for in writing the *Ornithology*, Ray (who was living there then) says: 'The bird is easily taught to speak, and that very plainly. We ourselves have known many, which had learned to imitate mans voice, and speak articulately with that exactness, that they would pronounce whole sentences so like to human speech, that had you not seen the birds you would have sworn it had been man that spoke.'[3]

Following Francis's funeral, Ray devised a structure for the ornithology book. Not only did he have to read everything previously published on birds, he had to evaluate the material and compare the descriptions made by different authors and assess whether particular species were the same as or different from those he and Willughby had seen.

The existing literature was much more extensive than one might think, for in addition to the three standard bird texts by Conrad Gessner, Pierre Belon and Ulisse Aldrovandi, as well as William Turner's little book on the birds of Aristotle and Pliny, there were several volumes describing the birds of foreign parts. These included Georg Marcgraf and Willem Piso's book on the birds of Dutch Brazil; a volume on exotic birds and other animals of the Dutch colonies by the Flemish physician Charles de L'Escluse (also known as Clusius); and Francisco Hernandez's work on the wildlife of Mexico. All substantial tomes. There was also Giovanni Pietro Olina's *Uccelliera* on cage birds, and the attractively illustrated but otherwise uninformative book *De Avibus* by Jan Jonston – later described as 'a hack writer with no firsthand knowledge of birds'.[4] Regardless, they all had to be assessed.

Willughby had a large personal library, and he and Ray were extremely well read both inside and outside natural history. They were familiar, for example, with William Harvey's *On Generation*, the catalogue to the massive cabinet of the Danish physician Ole Worm (Wormius), *Museum Wormianum*, and Antonio Pigafetta's account of Ferdinand Magellan's epic expedition to circumnavigate the globe, all of which contained valuable oddments of ornithological information.[5]

Some of these volumes contained illustrations of birds, but Willughby and Ray also searched more widely to find what they considered the most accurate images of particular species. In the end, the *Ornithology* contained illustrations from a wide variety of published and unpublished sources. Among the former, those from Aldrovandi, Piso and Olina are most frequently used, but the published sources also include a few images from the Flemish engraver Adriaen Collaert, the Spanish Jesuit mystic Juan Eusebio Nieremberg, Clusius and Worm. Unpublished images included paintings they borrowed from their friends and Royal Society colleagues, including Philip Skippon, Thomas Browne and Walter Charleton; and, of course, Willughby had his own images, including those in Baldner's book, purchased while on his travels.[6]

John Ray incorporated advice, information and details of specimens that Willughby and he had acquired from their regular correspondents, including Martin Lister, Francis Jessop, Sir Thomas Browne, Sir Philip Skippon and Ralph Johnson. Their colleagues

Christopher Merret and Walter Charleton, both Fellows of the Royal Society, were busy conducting their own bird projects during the 1660s. Merret was a physician, and a 'difficult' man, who suffered from melancholy and depression. In 1666 he published an incomplete alphabetical list of 'all' known British birds, plants, butterflies and minerals, entitled *Pinax Rerum Naturalis Britannicarum* (*An index of British natural things*), which included some brief notes on a few species sent to him by Willughby.

Two years later, in 1668, Walter Charleton published his list of birds and other animals, in a book entitled *Onomasticon Zoicon plerorumque animalium differentias* (*A zoological glossary expounding the distinguishing features and proper names of most animals*), the birds largely comprising those mentioned in the works of other authors and the birds present in the Royal menagerie in St James's Park, London, and in the Royal Society's museum. The *Onomasticon* included six lifesize monochrome etchings of birds on folding pages, and all quite well executed – a hawfinch, crossbill, 'benefico' (a small, unidentified songbird), bee-eater, hoopoe (a rare vagrant to Britain, killed near London!) and a pin-tailed sandgrouse, all based on paintings by an unknown artist owned by the English politician, Sir Thomas Crew. Neither Ray, nor presumably Francis Willughby, thought much of either Merret or Charleton's works, and curiously, to my mind, resisted the temptation of recycling any of Charleton's bird pictures in the *Ornithology*.[7]

Compiling and organising all this information into a seamless whole was a tremendous task for Ray. In contrast, putting quill to paper and writing the text must have seemed relatively easy, even though the book would eventually comprise one-third of a million words.

🐦

The cornerstone of the *Ornithology* was the 'arrangement' (or classification) of birds. As John Wilkins had made clear to Willughby and Ray soon after they met in 1660, classification was the key to 'organised knowledge' – of anything, whether it be words and games,

A hoopoe from Charleton's Onomasticon *(1668).*

or animals and plants – and part of the new science. With a reliable classification of birds, you could build on that to flesh out the biology.

From early on in their partnership Willughby and Ray recognised that the unambiguous identification of birds (and other organisms) was essential and could be attained only by developing and employing a systematic methodology. A system was vital to ensure accurate and consistent descriptions of both the outside and inside of birds. Armed with good descriptions, which provided a reasonably clear idea of the boundaries between species, they were in a strong position to evaluate the similarities and differences between different birds and create a classification.

Previous attempts at bird classification by the likes of Belon and Aldrovandi suffered from several shortcomings. Their lack of a consistent methodology, for example, meant that much of their identification was ambiguous, resulting in the 'multiplication' of species. Add to this their poor choice of classificatory criteria and here was an altogether shaky foundation on which to build.

As the new science unfolded, the need for a robust classificatory system became increasingly obvious and several of Willughby and Ray's contemporaries were tempted to try their luck. Walter Charleton's arrangement of birds, first discussed at the Royal Society in 1662 and published in 1668, serves as a model for everything that Willughby and Ray felt was wrong with natural history. Charleton, educated at Oxford and tutored by John Wilkins, trained as a physician but ended up more of a writer than a medical practitioner. He was elected to the Royal Society in 1662, yet the new science seems to have passed him by and his classification of birds is distinctly Aristotelian in its approach. The ornithologist Erwin Stresemann later said of Charleton's classification: 'nothing could more clearly demonstrate the uselessness of supposedly Aristotelian principles for avian classification than [this] which defies common sense'.[8]

In contrast, the scheme created by another colleague, Ralph Johnson, a clergyman-naturalist based in Brignall in North Yorkshire, was apparently much better. Ray says that both he and Willughby were 'informed' by it, suggesting that Johnson had some good ideas. They were also reassured by his arrangement of birds since it confirmed many of their own suggestions. Sadly, we know nothing of Johnson's scheme.[9]

The basis for Willughby and Ray's arrangement of birds was anatomy: feet and beaks principally. Centuries earlier Pliny the Elder had pointed out that 'The first and most important distinguishing characteristic among birds is the feet.' In 1575 the Dutch anatomist Volcher Coiter, one of Aldrovandi's pupils, also used anatomical features – mainly feet – to create a dichotomous classification of birds, but curiously, Willughby and Ray seem not to have known of his work. On the other hand, they probably were familiar with two other sixteenth-century writers, the anti-Aristotelian scholar Pierre de La Ramée and the travel writer Theodor Zwinger, both of whom popularised the idea of dichotomous keys.[10]

Armed with their detailed descriptions, Willughby and Ray started their new classification, dividing birds initially into landfowl and waterfowl, as several of their predecessors had done, but then – and this was the novel aspect – using details of beaks and feet and a few other features, such as body size, to create successive subdivisions.

Landfowl, for example, were divided into those with either (i) curved beaks, or (ii) straight beaks and claws. Then, for the latter group, these were divided into three categories based on body size: large, medium and small birds. Large birds were very large and included the ostrich, cassowary and dodo. Middle-sized birds were further divided according to whether their beak was 'large, thick and strong' (crows and woodpeckers) or 'small and short' (pigeons, poultry and thrushes). The birds in the latter category were further divided by the colour of their flesh: birds with light flesh were poultry (a feature that Willughby and Ray felt was unique); birds with dark flesh were either large ('pigeon-kind') or small ('thrush-kind').[11]

Willughby and Ray recognised that the function of such branching keys was twofold. The first was to reveal the 'true' arrangement of animals and plants, as devised by God and invariably elusive because it wasn't always obvious that God wanted anyone to fully know His plan. The second was an arrangement whose aim was to aid identification.[12]

It is this second type of arrangement that is included in the *Ornithology*, as Ray explains in the Preface:

By so accurately describing each kind [of bird], and observing their characteristic and distinctive notes [features] that the reader might be sure of our meaning [i.e. which species] and upon

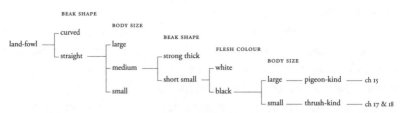

Part of Willughby and Ray's classification, redrawn from the Ornithology, *'A table of land-fowl', p. 54, showing a section of their not quite dichotomous key (note the three options, rather than two, for body size on the left). Criteria are in uppercase. The final parts on the right-hand side, for example 'thrush-kind', direct the reader to the relevant chapters where they can read individual descriptions to establish which of the various thrush-like birds they have in front of them.*

comparing any bird with our description not fail of discerning whether it be described or no.[13]

In other words, by using the *Ornithology*'s detailed descriptions, a reader should be able to identify any particular species, or establish whether it has previously been described or not. Ray continues:

Nor will it be difficult to find out any unknown bird that shall be offered: for comparing it with the tables first, the characteristic notes [features] of the genus's from the highest or first downward will easily guide him to the lowest genus; among the species whereof, being not many, by comparing it also with the several descriptions the bird may soon be found.[14]

What he means here is that if someone finds an unfamiliar bird specimen, then using the tables (the keys) the individual will be led to smaller and smaller groupings – as Ray says – until these are sufficiently small that it won't be too onerous for that person to read the species descriptions in order to match the specimen with a particular species.

What I don't think Willughby and Ray's arrangement allowed one to do – because this is an identification guide rather than a 'true' classification – was to classify a new species.

On the other hand, when I tried their scheme with a European bird that Willughby and Ray did not really know – the pin-tailed sandgrouse (see page 142) – I was surprised by the outcome. A land bird, obviously, and with a short beak, short claws (clearly not talons), and middle sized, but then I'm stuck because I don't know what colour its flesh is. Regardless of whether the flesh is pale or dark, Willughby and Ray's table suggests that the sandgrouse is either of the poultry-kind or pigeon-kind. This is pretty remarkable given that subsequent ornithologists classified sandgrouse first as a kind of grouse (i.e. closely related to poultry) and later as a kind of pigeon (because of the sandgrouse's unusual ability to drink by sucking, a trait shared with pigeons).

The reason that Willughby and Ray failed to include the sand-grouse in the ornithology was because they had only a painting to go on, and they didn't trust the artist's depiction. I suspect that had they had a specimen in front of them, they would have done at least as well as some later taxonomists in considering it some kind of poultry or pigeon.

The arrangement that Willughby and Ray eventually came up with is both ingenious and effective. The secret of its success was in finding those criteria that worked. There were so many possibilities – quantifying the number of flight feathers, the size of the gizzard, and the number of liver lobes – that finding the right ones was a stroke of genius. It was for this reason that subsequent authors revered Willughby and Ray. Indeed, when Carl Linnaeus, who is much better known than either of them, came to write his *Systema Naturae*, he failed (or refused – arrogant man that he was) to rec-ognise that the arrangement created by his two predecessors was superior to his own.[15]

Almost a year after Francis's death John Ray married Margaret Oakley, governess of the Willughby children at Middleton. Ray was forty-four and Margaret twenty. Their courtship had started before Willughby died, but completely unaware of it, Francis had said in a conversation with Emma that he didn't think Ray would ever marry; instead, he thought Ray would continue to live at Middleton, or wherever Emma chose to reside, and take care of her and educate their children. When Emma suggested to Francis that Ray might marry their sons' maid, he 'smiled at the conceipt & would not beleeve it'.[16]

The marriage – conducted by the family's curate, George Antrobus, on 5 June 1673 – began badly as Ray was unable to fulfil his connubial duties. Suffering from both erectile dysfunction and premature ejaculation, he sought advice from Martin Lister. He in turn prescribed *Diasatyria* – crushed 'Spanish fly' (actually a beetle, or one of several species known as blister beetles). Its key constitu-ent, cantharidin, acts – at an appropriate dose – as a gentle irritant on the genital and urinary tract, thereby increasing blood flow and facilitating erection. Transcribing Ray's correspondence after his death, his friend William Derham destroyed most of the letters dealing with Ray's sexual dysfunction, and it was only through the careful research by the historian Anna Marie Roos that Ray's correspondence with Lister was subsequently discovered.[17]

Cantharidin has been known as a sexual stimulant since ancient times. But it was also recognised that higher doses could be lethal. In 1772 the Marquis de Sade almost killed two prostitutes at one of his orgies by giving them aniseed-flavoured sweets laced with too much cantharidin.[18]

Nine years before Francis's death, in 1663 while in Modena, northern Italy, Willughby and Ray had come across a great bustard for sale in the market. This enormous bird – among the largest of all flying birds – still bred in Cambridgeshire and other parts of England at that time, but they were unaware that it also occurred in Italy. Suspended by its neck, the bird's gorgeous russet and white plumage seemed more like fine fabric than feathers. What neither Willughby nor Ray could have known was that the males of this species 'self-medicate' with blister beetles during the breeding season, presumably to enhance their sexual performance. Male bustards, like the ruffs mentioned earlier, congregate and compete at special display grounds each spring to secure copulations from passing females. In their efforts to impress, the males seem to turn themselves inside out in a most unbird-like feathery extravaganza. As with all lekking birds, the competition between males is so intense it is hardly surprising that some are tempted to enhance their performance with dangerous substances.[19]

The blister beetles must also have worked for Ray since he eventually fathered twin daughters Margaret and Mary in 1682, and a third daughter, Catherine, in 1687. Later, when discussing reproduction in *The Wisdom of God*, Ray asked, rhetorically perhaps, why there should be 'implanted in each sex such a vehement and inexpugnable appetite for copulation?' I wonder whether this particular phrase was prompted either by his frustration at not being able, at least initially, to make the most of his marriage, or possibly by the consequence of Lister's cure. In the light of natural selection the answer to Ray's query seems obvious, but in those pre-Darwinian days it was much more than a rhetorical question.

Overwhelmed by the family's expectations of him and by the effort needed to bring the bird book to fruition, Ray wrote to Peter Courthope in January 1674 saying that 'The death of Mr Willughby hath cast more businesse upon me than I would willingly have undertooke.'[20]

Two events that followed in rapid succession precipitated more change and further difficulties. In July 1675, Willughby's mother, the seventy-year-old Middleton matriarch Lady Cassandra, died. Ray had known for some time that without her backing his position in the Middleton household would be precarious at best. In a letter to Peter Courthope he had said that he would probably remain at Middleton 'at least so long as my old lady lives'. And certainly, soon after her death he and Margaret moved out, setting up home first at Coleshill (about seven miles south of Middleton) and in April the following year at Sutton Coldfield (some three miles from Middleton). Then, in August 1676, after Willughby's widow married Josiah Child, governor of the East India Company, she and the children left Middleton to live with him at Wanstead Manor, in Essex, a property so magnificent it became known as the English Versailles.

In May of that year Emma's sister-in-law Lettice had urged her to consider marrying again, writing to say: 'being in ye floure [flower] of your age tis pitty if God pleased you should passé ye remainder of your days in solitude'. However, by that time Emma's father Henry Barnard may have already arranged the union with Child, motivated to make the Middleton estates financially secure. At the same time, Barnard acknowledged that caring for the children from Josiah Child's two previous marriages might not be easy for Emma. Although her marriage to Child did secure the financial future of the Middleton estate – he was an avaricious and ruthless businessman – it came at an enormous cost to Emma, to the Willughby children, and to John Ray himself. Josiah Child was the polar opposite of Francis in every respect and soon proved himself to be a less than ideal husband and stepfather.

Apart from the soured relationship between Emma and Ray and the fact that her new husband also disliked him, the closure of Middleton meant that Ray no longer had access to Willughby's books and papers. This probably mattered relatively little for the ornithology project that Ray had completed by December 1674, but for the fish and insect projects it was to prove a serious impediment.[21]

Just before Christmas 1675 the printers sent Ray copies of the *Ornithology*, whose official publication date was 1676. Written in Latin, its title celebrated Ray's patron and collaborator: *Francisci Willughbeii: Ornithologiae Libri Tres*. Spectacular in its scope, this was also the most 'scientific' book on natural history so far, with none of the hieroglyphs, pandects (lengthy treatises) and emblematics (moral guidance) that cluttered the pages of most previous books on birds.

The writing was by Ray, who had used both Willughby's notes – now lost – as well as his own. As Ray says in the Preface, their aim was to illustrate the *history* of birds, by which he meant the results of their own investigations.[22] The *Ornithology* was designed as a system for understanding how different species of birds are related to each other: their natural order. This in turn depended crucially on correct identification, which in turn was based on the 'distinguishing marks' of each species.

Ray at least believed that God in His wisdom had created a fixed (albeit unknown) number of species.[23] The identification of some birds, such as the bullfinch, for example, was straightforward: the male's breast being a vivid pink, while that of the female is a gentle puce, and that of the immature a rich russet; and crucially all of them distinct from other finch species. Some birds, however, required more effort, as Ray relates with regard to the jack snipe: 'I sometimes following the vulgar [common] error, thought it not to differ from the Snipe [common snipe] in kind, but only

in sex, taking it to be the cock-snipe. But after being advised by
Mr. M. Lister, I found it to differ specifically: for dissecting several
of these small ones [i.e. Jack snipe] some proved to be males, some
females.'[24]

Ray and Willughby differed in the way they thought information should best be conveyed, with Ray inclined towards written
descriptions and Willughby more in favour of illustrations. Ray,
after all, had relied on careful, often vivid descriptions of plants
in his previous books – not least because he was unable to afford
engravings. Willughby on the other hand was wealthy enough to
purchase images and had clearly anticipated using those he bought
as the bases for illustrations in the *Ornithology*. However, it simply
was not possible to depict the distinguishing marks of each species in the illustrations, and these are restricted to written descriptions. In some cases, of course, they are there if you scrutinise the
images closely enough, but in contrast with some contemporary
field guides, they are not obvious.

For Ray, written descriptions were hugely important. When
Henry Oldenburg offered to have engravings made of some exotic
birds as the *Ornithology* was being prepared in 1674, Ray responded
by first accepting Oldenburg's offer, but also adding that:

> I must entreat some friend to take description of them in words,
> I mean their bigness, shape of the whole body, & particularly of
> their bills, feet and claws, colour of their bills legs and feathers,
> especially of their wings & tails, the length & figure of their tails
> & any other considerable or distinctive accident.[25]

This is a telling comment, for contrary to the popular saying, a
picture is not always worth a thousand words. An image – and
especially a seventeenth-century black and white engraving – con-
veys only so much. It cannot convey any aesthetic sense of the
bird; nor can it communicate overall body size or be sufficiently
detailed to show Willughby's distinguishing marks, such as the
exact number of feathers in the wing and tail. The truth is that
written descriptions and images *together* provided the material

necessary to provoke the reader's imagination and create a mental likeness of a bird.

There is some extraordinary evidence for just how effective the *Ornithology's* written descriptions are. When John Ray was arranging the production of a coloured copy of the volume as a gift for Samuel Pepys, the (unknown) artist had to rely almost entirely on the text, both for the colours themselves, but also for the region of the bird's body to which they were applied. Remarkably, in all but a couple of instances, this was done quite accurately.[26]

As well as detailed written descriptions, Willughby and Ray both recognised that the identification of species was enhanced by good illustrations. It was precisely for this reason that Willughby had so assiduously sought, and where necessary purchased or commissioned, colour images of birds during their travels. Many of those they acquired – such as Leonhard Baldner's book of paintings – were used in the production of the engravings. In other cases, they used images from earlier publications, some of which were excellent, like the finches and larks in Olina's *Uccelliera*.[27] On the other hand, they sometimes had no option but to use very poor pictures, such as those in Georg Marcgraf and Willen Piso's book of Brazilian birds.[28] The birds in that volume were unknown outside South America and their Portuguese names – such as *Pitangaguacul* or *Yzquauhtli* – were incomprehensible and unpronounceable. But because their goal was to describe all known birds, Willughby and Ray were compelled to include those images even though they were unhelpful and the names unfathomable.[29]

At least three of the images included in the *Ornithology* are of exotic birds that were, or had been, on display in St James's Park, London. As somewhere to walk with his dogs, courtiers and mistresses, Charles II had had the park restored and restocked with 'An abundance of fowl, and of several sorts, viz., both of land and water, viz., cranes, storks, shovelars, pelicans, ets ... Peacocks, peahens, a white raven ... partridges ... outlandish geese, ducks of several shapes, collours and sizes.' Willughby and Ray visited the park sometime between the late 1660s and early 1670s and were undoubtedly thrilled by the various unusual species on view, and by three in particular: the ostrich, cassowary and an Egyptian goose. It seems that the king commissioned Francis Barlow, already fashionable among the aristocracy for his depictions of birds and other animals, to create lifesized oil paintings of these particular species. It was from these paintings that the engravings used in the *Ornithology* were made. Frustratingly, we do not know how this came about: were Willughby and Ray invited to see the paintings and did they then seek permission from the king to have engravings made? Rather surprisingly, Ray is silent on this point in the *Ornithology*.[30]

It was sometimes hard for Willughby and Ray to know whether a description in the works of other authors referred to a real bird or to some mythological beast. Ingeniously, Ray dealt with this by including an appendix of what he referred to as 'Such birds as we suspect for fabulous'. Among these were some fantasy species, such as the phoenix. But there were others that subsequently turned out to be genuine, such as the hoatzin that Ray correctly describes as living in hot countries and 'very often is found sitting in trees by rivers'; and less accurately, that it 'feeds upon snakes' (it is vegetarian), and 'its bones asswage the pain of any part of a man's body by launcing'. Untested as far as I know, but unlikely.[31]

Another problematical bird was one mentioned originally by Antonio Pigafetta, who as noted had travelled with Magellan to the Philippines and East Indies in the early 1500s. The 'daie' was said to lay a great abundance of enormous eggs that were buried in 'deep vaults within the ground' and hatched without ever being incubated. Because this seemed stranger than fiction, Ray rejected it, writing: 'I dare boldly say that this history [account] is altogether fabulous.' He was wrong, for the daie is the Philippine megapode, one of several species of large-footed birds that dig cavities in either warm volcanic soil or mounds of rotting vegetation to deposit and

Red-backed shrike showing the tomial tooth (notch) towards the tip of the upper mandible, noticed by Willughby (top); nestling woodpigeons (bottom) – Willughby speculated on whether pigeon broods (always two) comprised one of each sex

European roller. Willughby identified several 'distinguishing marks': the tongue (top); the wart-like region behind the eye (middle); and the structure of the toes

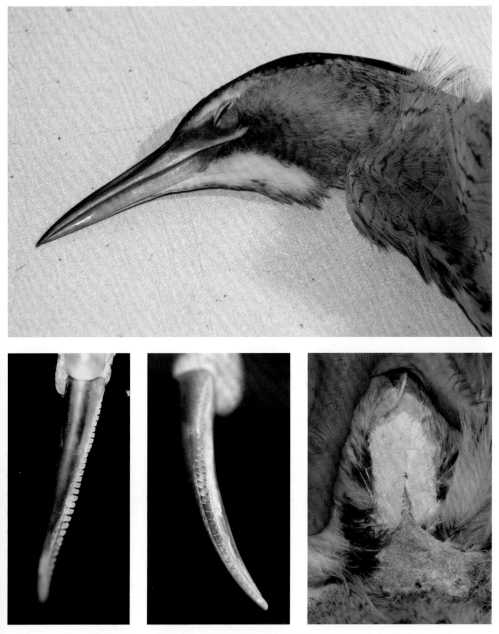

Eurasian bittern. Willughby and colleagues dissected one while he was an undergraduate at Cambridge: head (top); pectinated middle claw (bottom left and centre) and the power-down feathers on the breast

The 'pigo', a roach-like fish whose spiky breeding-season scales captured Willughby's imagination

TAB. XLIIII.

Coccothraustes.
The Grosbeak or Haw-
finch.

Coccothraustes Indica
cristata.
The Virginian Nightingale.

Chloris.
The Greenfinch.

Rubicilla.
The Bullfinch.

Loxia.
The Crosse-bill.

Passer.
A Sparrow.

Plate from the hand-coloured edition of *The Ornithology of Francis Willughby*, owned by Samuel Pepys. Clockwise from top left: hawfinch, northern cardinal (a North American import), bullfinch, house sparrow, crossbill and greenfinch

Title page of Leonard Baldner's unpublished book, purchased by Francis Willughby while in Strasbourg; note the portrait of Balder (right)

The snap-apple, as the common crossbill was known in Willughby's day: male (top) and female (bottom)

John Ray by William Faithorne, around 1690 (top left); John Wilkins (top right); and Wollaton Hall, Nottingham, built by Francis Willughby's great-grandfather, Sir Francis Willoughby

hatch their eggs. Given what was known about birds in the mid-1600s, and the extraordinary biology of these and other megapodes, Ray and Willughby were right to be cautious.

The *Ornithology* is, as indicated by its Latin title – *Libri Tres* – made up of three books or parts. These are preceded by a Preface, in which Ray describes how the book came about and celebrates Francis Willughby. Book one is an overview of bird biology – essentially, what birds are and how they work – spanning sixteen pages. This is followed by the account of some 'English' seabird colonies – and Francis Willughby's questions that we discussed in the preceding chapter. These pages are followed by their classification, and a catalogue (list) of English birds. Books two and three cover Willughby and Ray's two major divisions of birds: land birds and waterfowl, species by species.

The final decision about what to include and what to exclude from the *Ornithology* lay with Ray. He states in the Preface that they present only what properly relates to natural history and included only material they can 'warrant upon our own knowledge' and 'what we approve'. It is rather surprising, then, to find within the book's pages some comments derived from ancient Greek and Roman authors on the use of birds in medicine, including swallows as a cure for 'falling sickness' or epilepsy.[32] On the other hand, the presence of this kind of information provides a revealing picture of the state of medicine in Willughby's day – certainly, there is no hint of scepticism. For example, Ray regurgitates Galen's notion that a crested lark roasted or boiled will assuage colic pains,[33] and Aetius's assertion that the wren is the 'perfect cure' for kidney or bladder stones 'being salted and eaten raw, or being burnt in a pot ... and the ashes of one whole bird taken at once'.[34] Similar curative properties are ascribed to the pied wagtail, but Ray refers to Alexander Benedictus who thinks that 'modern physicians' may have muddled wagtails and wrens, adding that Conrad Gessner thought 'it matters not much what birds be burnt, sith [since] the vertue of the ashes of almost all birds seem to be the same'. I suspect Gessner was correct.

Ray is more sceptical about non-medical folklore involving birds, relating the 'vulgar persuasion' (common belief) that the kingfisher 'being hung up on an untwined thread by the bill in any room, will turn its breast to that quarter of the Heaven whence the wind blows'. He then adds – sarcastically, I think – 'He that doubt of it may try it.'[35]

The *Ornithology* is a blockbuster, a massive compendium of ornithological knowledge. It is indicative of Ray's organisational genius that he presents the information in a way that has survived the passage of time, for the book's layout provided a model for virtually all subsequent bird encyclopedias.

Despite Willughby and Ray's excellent 'arrangement' of birds, the classification of animals – including birds – continued to be, and continues to be, a monumental issue for zoologists. In the 350 years since Willughby and Ray, classification has continued to preoccupy ornithology, and during the eighteenth and nineteenth centuries the study of birds consisted of little else. There was a brief hope that Darwin's theory of evolution by natural selection in the mid-1800s would provide the key to unlock the secret, but avian classification continued to confound even the most dedicated ornithologists. Classification became the domain of museum men measuring and perusing the preserved skins and skeletons of birds from around the world.[36] Few bothered to look at how birds behaved or what they did in the wild – because it was felt that behaviour or ecology would add little or nothing to their ongoing classificatory campaign. It was not until the early twentieth century that behaviour was considered a worthwhile taxonomic criterion, and not until the early twenty-first century that molecular tools were sufficiently honed that they could be employed to provide a truly objective – and what we now assume to be a fairly accurate – picture of how different groups of birds evolved and are related to each other. Interestingly, despite its initial promise, molecular biology has not yet provided *all* the answers because some groups of birds, such as gulls for example, diversified so rapidly in their history that the differences in their DNA, examined so far, are too small to unambiguously distinguish their evolutionary pathways.[37]

A page from Pierre Belon's (1555) bird book, L'Histoire de la Nature des Oyseaux, *showing the text and woodcut together.*

At the very end of the *Ornithology* are eighty plates, depicting 380 individual birds. Their inconvenient location at the back of the book, separated from their relevant text, is a consequence of metal engraving. The woodcuts used previously by the likes of Conrad Gessner and Pierre Belon permitted printers to place images and text on the same page, but those images were very crude compared with what could be achieved by metal engraving.

The plates for the *Ornithology* were produced by William Faithorne (who also created an excellent portrait of John Ray), William Sherwin and Frederick van Howe, London engravers employed by the Royal Society's printer John Martyn. The results were mixed, with some more lifelike than others, reflecting a combination of what was often poor material to start with, a dearth of ornithological experience among the engravers, and the lack of anyone to oversee their production. Ray was disappointed, and says so in the Preface:

> The gravers we employed, though they were very good workmen, yet in many sculps they have not satisfied me. For I being a great distance from London, and all advices and directions necessarily passing by letter, sometimes through haste mistook my directions, sometimes through weariness and impatience of long writing sent not so clear and full instructions as was requisite, as they often neglected their instructions, or mistook my meaning.[38]

Unless you know someone who has been involved in producing a copiously illustrated book, even with the advantage of today's technology it is difficult to imagine just how much effort Ray must have expended in organising the images for their book. Little wonder then that there are a few glitches, with one or two species, such as the blackbird and Royston (hooded) crow, appearing twice, and some species curiously placed, as with the yellow wagtail among the gulls and terns.[39]

Just reading and checking the proofs of over 300 pages of text must have also been an enormous task. But Ray was an efficient, systematic worker, and I can easily see him sitting down with a pile of page proofs and methodically going through them to check for errors. He did so with meticulous care, for the finished book contains very few mistakes.

Plate 40 from the Ornithology *showing (from top to bottom): yellowhammer* Emberiza citrinella, *corn bunting* E. calandra, *ortolan bunting* E. hortulana, *skylark* Alauda arvensis, *woodlark* Lullula arborea, *crested lark* Galerida cristata *(bottom left), all redrawn from Olina (1622). Interestingly, the bird Willughby and Ray thought was a yellowhammer in Olina, and the one labelled as such in their* Ornithology *(top right here), is actually a cirl bunting* E. cirlus, *as Olina's text makes clear: 'its head is the colour of a serin, greenish-yellow'.*

As soon as Ray received copies of the *Ornithology* from the print-ers, he sent one to Martin Lister to thank him for his help. Lister responded by writing to say how pleased he was to have it, but that Ray should produce an edition in English, because he felt (correctly as it happened) that the Latin version would have limited appeal. Lister also suggested some additions that he thought would make the book more attractive to the book-buying public, including a section on hawks (and falconry) and 'some account of the keeping and ordering of [song] birds in cages'. For the latter he recom-mended two recent books from which Ray might glean the appro-priate information: *The Gentleman's Recreation* (1674) by Nicholas Cox and *Epitome of the Art of Husbandry* (1669) by Joseph Blagrave. Ray ordered the books and read them, but he was unimpressed. He wrote back to Lister on 4 April 1676 pointing out in his perceptive way that Cox's book was simply a plagiarised version of *The Art of Fowling* published in 1621 by Gervase Markham, and that most of the material in Blagrave's book was similarly lifted, from the works by Aldrovandi and Olina. The only section in *Epitome* that Ray thought might be original referred to the breeding of canaries that were just starting to become popular as cage birds in Britain and on the continent. Regarding the plagiarists, Lister responded: 'This sort of men being the bane and pest of learning, and you ought to brand them.'[40]

Lister also recommended including an account of Faeroese sea-birds by the Danish priest, Lucas Jacobsen Debes. This was pub-lished initially in Danish in 1673, but as Lister knew it had just, in 1676, appeared in an English translation. At the time Ray was writ-ing, there were no copyright laws and nothing to protect authors from plagiarism, and Ray reproduced Debes's account verbatim, but at least he acknowledged its author.

That Ray agreed so readily to Lister's suggestions may have been because he held him in such high regard and recognised his sound judgement, but also because for his own financial security he needed the *Ornithology* to be a commercial success. Ray may also have acquiesced to Lister's suggestions as a goodwill gesture to make amends for a disagreement they had had a few years earlier. That disagreement was significant not just at a personal level, but because it marked a change in the way practitioners of the new science dealt with discoveries. The dispute concerned the ballooning behaviour of spiders.

On the roof terrace of a tall house in Spain the mountainside opposite me lies in deep shade. The late afternoon sun lights up a billion airborne flecks in the same way a projector lamp lights dust in a room. The warm breeze is blowing insects, other invertebrates and the feathery seeds of numerous plants horizontally along the mountainside. In among this mass of aerial plankton are also the diaphanous threads of hundreds of strands of spider silk. Unlike the horizontal trajectories of the seeds and insects, the silken threads hang almost vertically, gently undulating and only occasionally catching the light and becoming visible. I initially imagine the threads to be just a metre or so long, but as I watch, I realise that some are enormous – over ten metres in length.

More than three centuries before, Martin Lister had seen something similar:

> [I]n close attending on one [spider] ... I saw her suddenly ... turning her tail into the wind, to dart out a thread with the same violence that water spouts out of a spring; this thread, taken up by the wind, was in a moment emitted some fathoms long, still issuing out of the belly of the animal; by and by the spider leaped into the air and the thread mounted her up swiftly.[41]

This is how spiders reach new habitats, and under certain weather conditions at particular times of year the air seems to be filled with their diaphanous linear parachutes. It is a sight as intriguing and beautiful today as it was when Lister first noticed it. Lister told Ray of his discovery, who without the former's knowledge related it to others with the unfortunate result that Lister's finding was published anonymously in the Royal Society's *Transactions*.[42] Lister was understandably annoyed, but the situation was made worse when a certain Dr Edward Hulse – a friend of Ray's – capitalised on the circumstances by claiming the discovery as his own.[43,44] Priority was beginning to matter. Before the advent of the new science and scientific journals, who had discovered what was not really an issue, but as the scientific revolution rolled forward, priority became increasingly important as scientific endeavour and discovery began to serve self-advancement.

Acutely competitive, Martin Lister felt the lack of credit keenly, which is ironic because he seemed to have had little compunction about capitalising on the unpublished findings of his friends. On hearing that Willughby and Ray were studying sap and insect reproduction, for example, he immediately began his own research on those topics without crediting either of them. Even so, in the case of the aerial spiders, Ray was in the wrong and keen to make amends.

Taking Lister's advice, Ray immediately began working on an English edition of the *Ornithology*, using the opportunity to make a few improvements to the Latin text. He included a more extended eulogy about Willughby; he identified those species that neither he nor Willughby had dissected themselves with an asterisk in the margin so it was clear that the description was based on the work of others. Ray also made an effort to make the text more user-friendly, to the extent of adding a little joke about a white blackbird, if the reader would 'pardon the seeming contradiction'. The English edition also included two additional plates to complement the new text on fowling: one comprising eight separate images taken from Olina's *Uccelliera* and the other consisting of two images from Markham's *The Art of Fowling*.

Ray also added a paragraph to the Preface to mention a change he considered making to the English edition, but didn't. This paragraph is a brief comment on the 'wholesomeness' of the flesh of birds as part of our diet. Pointing out that the individual species accounts include some notes on the palatability of birds, Ray also states that, because many authors have included similar information in the 'dietical part of their institution of physic', he provides only a summary. It consists of four points:

1. The flesh of carnivorous birds is not worth eating; nor is that of 'the crow-kind'– that is, corvids, such as rooks, crows, jackdaws, jays and magpies, which feed 'promiscuously upon flesh, fruit or seeds'.
2. Birds that feed on insects, such as woodpeckers and swallows, are not good to eat either, although 'small birds of slender bills that are reputed good [to eat] ... feed as well on fruits and berries ... but are best when they feed upon fruits, as the Beccafigo [blackcap or other warblers] in fig-time', which means when they are fattening up for migration.
3. 'The birds that feed upon grain and seed only ... as the poultry kind, are best of all.'
4. 'Waterfowl, such as feed only or chiefly on fish are not good meat; yet the young of some of these are approved as a delicacy, though I scarce think wholesome: such are young soland-geese [gannet], puffins [Manx shearwaters rather than Atlantic puffins, I think], pewets [black-headed gulls] and herons.'

Ray then says that waterfowl, ducks and geese, although feeding on insects, are 'esteemed good to eat and admitted to our tables', adding that 'The most delicate of these are those we have termed mud-suckers, that with their long nebs [beaks] thrust into the earth suck out of the mud or ouse [ooze] a fatty juice, by which they are nourished.' A 'fatty juice' sounds remarkably unspecific. Further on in the main body of the *Ornithology*'s text Ray elaborates slightly by defining mud-suckers as having very long bills, either 'crooked' (curved) as the curlew, or 'streight' as the woodcock, 'which suck out of the mud or channels some oyly [oily] slime or juyce, wherewith they are nourished: whence they have delicate flesh, and their very guts not emptied or cleaned from [of] the excrements are usually eaten'. In case this is not clear, Ray is referring here to the habit of cooking these mud-sucking birds without removing or cleaning the guts, and indeed, eating the guts along with the breast and leg muscles. This is a habit that still persists today with those that eat woodcock and snipe: oh yes, and although Ray doesn't mention this, the head is left on, so that the brain can be eaten too.[45]

In the individual species accounts, Ray states that birds such as woodcock, curlew and snipe eat worms and insects, but I presume the reason for his assuming that the main diet of mud-suckers consists of 'oyly slime' is because the soft-bodied worms that form the main part of their diet are rapidly digested and leave little identifiable trace within the gut when dissected. Closely linked to this is the widespread belief that when a woodcock or a snipe is flushed from the ground (before being shot), it defecates, and for these reasons it is acceptable to eat their guts.

Guts apart, the woodcock, along with the grey partridge, was among the most highly esteemed table birds. Ray writes:

> The leg especially is commended, in respect whereof the woodcock is preferred before the partridge itself, according to that English rhythm [rhyme]:
>
> > If the Partridge had the Woodcocks thigh
> > Twould be the best bird that ever did fly.[46]

One reason, I think, why Ray includes this brief account of which birds are worth eating or not is because during his and Willughby's continental travels they were often amazed by the birds – including hawks, corvids and starlings – that local people were prepared to eat, that no one in England, at least no one that Ray knew, would consider eating. The explanation for this geographical variation in what birds were considered palatable may simply have been a difference in tradition, or necessity. That people in Britain did not typically eat raptors, corvids or starlings is confirmed by Muffet's book *Health's Improvement*: raptors get no mention at all, and both corvids and starlings are not recommended as worth eating, except in the case of corvids, as nestlings.[47]

The English version, entitled *The Ornithology of Francis Willughby*, was published in 1678 and, in contrast to the Latin edition, made no mention of Emma as the sponsor of the engravings. This isn't too surprising, for by this time relations between her and Ray had degenerated to such an extent that he felt under no obligation to acknowledge her. We don't know whether the English edition yielded the economic return Ray hoped for during his own lifetime, or indeed whether Emma benefited financially. But in terms of its impact on the study of birds the *Ornithology* was an unparalleled success.

At Wanstead meanwhile the situation between Josiah Child and his stepson Francis junior had deteriorated so much that, in June 1680, the twelve-year-old asked his uncle Lord Chandos, Francis Willughby's brother-in-law, if he could come to live with him. Chandos, however, was horrified at the prospect of having to care for young Francis and urged him to return to his mother. Instead, Francis junior ran away from Wanstead to his aunt, Francis's sister Lettice Wendy, who lived at Haslingfield in Cambridgeshire. Dismayed by the boy's sudden appearance, she nonetheless agreed to take him in.

Francis junior's unhappiness was largely down to his relationship with his unpleasant stepfather, and his escape from Wanstead may have been triggered by the death of his maternal grandfather, Henry Barnard, in April of that year. The departure of John Ray and Margaret Oakley, both of whom had taught the Willughby boys while at Middleton, must have further added to his distress. Josiah Child was clearly incapable of providing the love and care that the Willughby children had previously enjoyed, and Francis junior also made sure that his younger brother Thomas and sister Cassandra eventually left Wanstead as well.

It is a measure of just how 'sordidly avaricious' Josiah Child was, that at one point he calculated how much his stepchildren had cost

Plate 52 from the Ornithology *showing (not to scale): white stork* Ciconia ciconia *(upper left), black stork* C. nigra *(upper right), Eurasian spoonbill* Platalea leucorodia *(lower left), and Eurasian bittern* Botarus stellaris *(lower right).*

him from the day he married their mother, Emma. After Barnard's death, Child also expected Francis Willughby's executors to undertake more responsibility for the estate and the ongoing inheritance battle. He singled out Ray for a very personal attack, accusing him of dishonesty and insisting that he should live closer than Essex – where he and his family had moved after Ray's mother died in 1679 – to justify the annuity by fulfilling his duties in educating Willughby's children. Child also involved his wife in the attack on Ray. The issue was that Barnard had not identified a replacement for Sir Thomas Wendy as one of Willughby's executors, after Wendy's early death in November 1673. As an executor Ray felt that it was his role to nominate other executors. This did not go down well in the Child household, and Emma wrote an angry letter to Ray accusing him of 'insufficiency, laziness and dishonesty'. He responded by firmly defending his position, reminding Emma that Francis Willughby had provided the annuity as 'a free gift in token of his kindness and friendship', adding that while he was prepared to assist with estate matters, he did not feel that Francis had assumed he would do so without help. He also reasserted that he was happy to oversee the education of her and Francis's sons when they were older. In response, Child withheld Ray's annuity for two years.[48]

The atmosphere within the Child household was intolerable. Torn between loyalty to her children and to her 'petty and controlling' husband, Emma was distraught because her oldest son had run away to live elsewhere, and because her other two children, Thomas and Cassandra, were thinking of leaving too.

}⸙{

After moving back to the house in which he had grown up, Ray began work on the fish book, announcing to Martin Lister in December 1674 that: 'Having finished the *History of Birds* I am now beginning that of fishes'. Eighteen months later he wrote again, optimistic that *Fishes* would be finished within a year or two. It took eight years. Part of the delay was a waning enthusiasm for fish and a waxing of his own botanical books.[49]

As well as using the notes he and Willughby had accumulated during their travels, Ray continued to seek additional fish specimens and information from colleagues. His Cambridge friend Peter Dent, an apothecary and a physician, asked his local fishmonger to keep an eye open for unusual species, and, in February 1675, Dent was able to send Ray a ten-pound thornback ray. He also told Ray how this same fishmonger had once sold a skate (or 'flair' as he called it) weighing over 200 pounds to the cook at St John's College, Cambridge, where it fed all thirty scholars. Ray chose to include this story in his *History of Fishes* where, as is so often the case with fishermen's tales, the flair had grown sufficiently large to feed 120 men.[50]

The book's formal title, *Historia Piscium*, reveals that like the first edition of the bird book it was written in Latin. Unlike the bird book, however, there was never an English edition, and apart from some small initial success, it has remained in relative obscurity ever since. I had some of it translated into English, and my overwhelming sensation is how glad I am that Ray himself translated the *Ornithology* into English. He knew exactly what he wanted to convey and translated the Latin with sufficient nuance that the English version reads beautifully and informatively. Had a non-ornithologist made the translation, I doubt whether the bird book would have been the success it is. By comparison, bits of the fish book are difficult reading, not because my translators didn't know their Latin, but because they couldn't know what it was that Ray wanted to say. It is analogous to a situation I sometimes experienced as a university teacher, when it was suggested that I deliver an undergraduate lecture based on someone else's notes (or vice versa): it never really works.

The fish book is organised in a similar way to the *Ornithology* in some respects, but it is very different in others. The overall aim was to provide a classification of fishes whose reliability was founded on accurate identification, which in turn was based on detailed physical descriptions and 'characteristic marks'. As Ray states: 'We particularly abhor the unnecessary multiplication of species'[51] – a problem prevalent across the whole of natural history because variation in morphology routinely resulted in the same species being known by different names and considered different species. As with the *Ornithology*, Willughby and Ray's account of the fishes excluded all myths and folklore and all those other tedious demonstrations of 'copious knowledge' favoured by their predecessors. The only concession was when folklore had some bearing on the biology of a particular species, such as the adage 'as dead as a herring', which accurately described how quickly that particular species expires out of water; or 'as hale as a roach' indicating the opposite.

The *History of Fishes* comprises four parts: (i) an introduction to the biology of fish; (ii) cetaceans (whales and dolphins); (iii) cartilaginous fishes (sharks and rays); (iv) bony fishes, making up the bulk of the book.[52] As Ray acknowledges, he was responsible for writing parts 1 and 2, and 'edited' the other two parts written largely by others. His structure, and hence classification, follows Aristotle rather than the more recent but less reliable Rondelet, who, as Ray says, sometimes separated species that are obviously closely related, such as splitting salmon from trout, and eels from congers.

The classification in the *History of Fishes* is based very clearly on what Willughby produced in 1666 for John Wilkins's *Real Character*, although it is also apparent that Ray was involved, for some of the fish mentioned are ones that only Ray had seen. Despite Wilkins's irrational and wholly artificial imposition of nine categories, that is what Willughby used for the fishes, and pretty much the system that Ray was to employ in the fish volume.[53]

Unless you are a fish enthusiast, and I suspect, even if you are one, the *History of Fishes* is hardly an exciting read. The book feels as though it has been written by committee, which of course it was. There are a few gems, however, and I particularly liked the account of the aptly named 'crampfish', that is, the electric ray or torpedo fish that Willughby and Ray encountered in Italy. Ray discusses its well-known 'narcotic powers', writing 'I handled, skinned and dissected several with my own hands and got no sort of harm: so it is evidently confined to the living fish.'[54] This is correct, but in life this species, which grows up to almost two metres in length, can deliver a powerful and debilitating electric shock.

Ray also describes the startling effect of a red sea-bream:

> We bought a large specimen and kept it overnight in the room meaning to dissect it in the morning: lo, in the dark the whole fish shone marvellously like a glowing coal or red-hot iron: never before had we seen such a phenomenon.[55]

This extraordinary sight was – we now know – the result of a very specific bioluminescent bacterium. Most bacteria that emit light produce a blue-green glow, but those in the genus *Vibrio* emit a beautiful yellow-orange light. It may even have been as a result of this glowing bream, mentioned perhaps in conversation, that Ray's fellow Royal Society member Robert Boyle later published an account of how, using his air pump to remove the oxygen from a container, he could extinguish the bioluminescence emanating from rotting fish.[56]

╠↑╣

Ray finished writing the *History of Fishes* in February 1685 and on 25 March informed the Royal Society that the text was complete. It had been assumed that the Society's printer, Mr Martyn, would publish the book, as with the *Ornithology*, but Martyn had just died so this was not an option. Dr John Fell, Bishop of Oxford, and Oxford University Press were approached and they agreed to publish it. Martin Lister, now living in London, was asked to take responsibility for the illustrations. Ray, having fulfilled his Willughby obligations and truly fed up with Emma and her husband, Josiah Child, dissociated himself from the publishing part of the process. He nonetheless suggested to the Royal Society that the expense of publishing might be offset by subscription, and failing that, perhaps Francis's son, now Sir Francis Willoughby, who was just sixteen in the spring of 1685, might find the funds when he came of age. In the same letter Ray mentioned that he already had some of the images, including some from Willughby himself, as well as those culled from other publications. However, the idea that not all the fish images were to be Willughby originals caused consternation among members of the Royal Society. Robert Plot wrote to Francis Aston, who was then secretary and responsible for the *Philosophical Transactions*, expressing concern that 'it had been presumed before, that Mr Willughby had taken all the draughts from … life, whereas it was now found, that the cuts [images] must be picked up here and there out of books'. Aston knew the power of high-quality, original illustrations for science and was worried that if some of the images in the fish book were not original, the entire exercise was at risk. It was for this reason that the Royal Society decided to oversee the production of *Historia Piscium*.[57]

Seeking subscriptions at a guinea a plate, the Royal Society found sponsors for all but thirty-three of the 187 plates. Seventy-four were covered by sixty-nine subscribers (many of them Fellows, including Christopher Wren, Robert Boyle, John Evelyn and Martin Lister), and no fewer than eighty by the Society's president, Samuel Pepys, who gave £50 initially and a further £13 later on. In response to Pepys's generosity, Ray dedicated *Historia Piscium* to him, but also arranged for Pepys to receive a hand-coloured edition of the *Ornithology*.

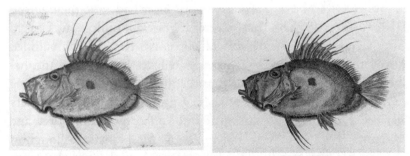

Illustrations of a John Dory (or St Peter's Fish) Zeus faber for Willughby's History of Fishes: *on the left the original watercolour; on the right the published engraving (Ray 1686).*

Unfortunately, the details of this gift are hazy. Although this coloured copy of the *Ornithology* bears both Samuel Pepys's coat of arms and his bookplate, apparently affirming its authenticity, no correspondence between Ray or Pepys has been found that mentions what must have been a spectacular and expensive present. This exquisitely coloured edition – now in the Blacker-Wood Library at McGill University in Canada – is thought to have been thrown out from Pepys's library after his death, as a 'duplicate', presumably by someone who hadn't bothered to look inside it.[58]

The fish book, notwithstanding its superb illustrations, was a financial disaster. It was too expensive; fish were less popular than birds; and political difficulties in the 1680s meant that people had more important things to worry about.[59] The fact that the *History of Fishes* did not sell as well as hoped brought the Royal Society to the very brink of financial ruin and meant, among other things, they were unable to meet the cost of publishing Isaac Newton's *Principia*. Fortunately, another Fellow, the wealthy astronomer Edmond Halley, came to their rescue.[60]

The final book produced by Ray based on Willughby's notes was about insects, *Historia Insectorum*, published in 1710 – five years after Ray's death. Thirty-six years earlier, on 30 November 1674, and two years after Willughby had died, John Ray had written to Henry Oldenburg, Secretary of the Royal Society, saying 'The History of Insects is that wherein Mr Willughby did chiefly labour and most considerably advance; which yet for some reasons I reserve for the last.' So even before the Latin edition of the *Ornithology* was published and before he had started on the *History of Fishes*, Ray knew that Willughby's work on insects was to be the last of his friend's books that he would bring to fruition.

Why did Ray produce the volume on insects after the other two books? One reason, it has been suggested, was that he wanted to avoid competing with Martin Lister's book *Historia Animalium* on snails and bivalves, published in 1678.[61] It may seem implausible that books on insects and molluscs would over-lap or compete, but in the 1600s the term 'insect' basically meant 'invertebrate' and included molluscs, as is nicely demonstrated by one of Lister's own papers in the *Philosophical Transactions* of 1678, entitled 'Tables of snails, together with some queries relat-ing to those insects'.[62]

I am not convinced by the idea that Ray postponed the insect volume to appease Martin Lister. Even though both books deal with 'insects' their contents are completely distinct. Lister's book is explicitly about molluscs, whereas Ray ignores these and focuses on other invertebrates – specifically worms, arachnids (spiders, ticks and scorpions), isopods (woodlice), as well as what he rather charmingly calls multipedes (meaning many legs: millipedes and centipedes) and, of course, insects proper.

There is another reason for thinking that the delay with the insect book had nothing to do with Lister's volume. Lister was well aware of the taxonomic extent and importance of Willughby's invertebrate research and indeed had written to the Royal Society in February 1682 to say that it would be 'a pity, that Mr Willughby's curious and voluminous observations on insects, in which he greatly delighted, should be lost'.[63]

The most likely cause, I think, for the delay, is that insects proved to be much more difficult to describe and classify than fishes, which in turn were more difficult than birds. Whatever the reason, Ray cannot have started on the insects until after the *History of Fishes* was completed in 1684. There remained a stumbling block, how-ever, since Willughby's notes were still held under lock and key by the irascible Josiah Child. The relations between Child and Ray's Royal Society friends continued to thwart efforts to gain access to Willughby's notes and specimens.

Things changed, however, in 1687, when Sir Francis Willoughby, Francis Willughby's son – the one who as a boy left home to live with his aunt – decided to move into Wollaton Hall, inviting his sister Cassandra to join him as housekeeper. Soon afterwards, they visited their old home at Middleton Hall 'to see in what condition the books, etc (left there by my father) were'.[64] This promising new chapter was thrown tragically off course when, after a brief illness, Sir Francis died the following year at the age of nineteen. Fortunately, his younger brother Thomas, to whom the baronetcy was transferred,[65] assumed responsibility for the family, moved in to Wollaton with his sister, and continued his brother's action against their mother's husband.

When Sir Thomas was still a boy, his stepfather Josiah Child had taken monies from the Middleton estate and invested them in the East India Company. Later, Child – again without any agreement from Sir Thomas – had allegedly sold timber from the Willoughby estate. And finally, Child refused to release capital that Sir Thomas needed to add lands to the estate. The case went to arbitration in 1689 and the amount owed to the Willoughbys was calculated – minus the amount that Child had spent maintaining Francis Willughby's children.[66] Once the case was settled the family was able to visit Middleton Hall and remove Francis Willughby's collections and books as well as some furniture, plate and linen. The full extent of what they recovered is not known, but Willughby's daughter Cassandra later commented on there being a collection of 'novell remarkables' (curiosities of some kinds), many books, and 'a fine collection of valuable medals, and other rarities which my father had collected together of dried birds, fish, insects, shells, seeds, minerals and plants and other rarities'. As this implies, as well as the medals and coins that Willughby obtained from the Dursley hoard a decade earlier, the collection must also have included the cabinet containing the extensive collection of seeds, fossils and birds' eggs, as well as Willughby's insect notes and specimens.[67]

Fortunately, we have a first-hand account of what the collection was like from James Petiver, who visited Wollaton for five days in the summer of 1691 as the guest of Sir Thomas and Cassandra. An apothecary, botanist and entomologist (his book on butterflies was one of the first), Petiver's visit was part of a two-month-long 'scientific peregrination' that started at his home in Kendall and ended in London. Francis Willughby's collection at Wollaton was one of the highlights of Petiver's trip and, as he noted in his diary, it sent him into an 'Extasie':

> I had never seen such a Collection of Naturall Things before (here are the Globes, the Spheres.) here are Trophies both from the East and west Indies, as well of Sea as the Lands, in Short it is a well furnished Museum, there are a great deal of Strang: Animals, some whole, and some in part, there is a good Collection of Coynes, and Medalls, of sea-shells, of Birdsegge shells, and the best Collection of Insects in England.[68]

A vast collection – and yet what remains (miraculously) in the seed cabinet today is clearly a fraction of what there once was.

Prior to Petiver's visit, and realising the significance of their father's work, Sir Thomas and Cassandra had begun the massive task of organising and preserving the collection. They were lucky to have the help of Dr Thomas Man, a keen botanist who had been Sir Thomas's tutor at Jesus College in Cambridge. Man also encouraged Sir Thomas's interest in natural history, which developed to such an extent that in 1693 he was elected to the Royal Society, 'unanimously chosen ... for his own qualifications as on the account of the great honour that the Society and everybody of it bears to the memory of yr father'.[69] It was undoubtedly Man's influence that inspired Sir Thomas and Cassandra to consolidate Francis's herbarium collection of plants – still in existence and currently housed at Nottingham University Library – assembled by their father when he and Ray had worked together. It must have been Thomas Man who encouraged Sir Thomas to collate and publish his father's insect material since Man prepared some engravings for it. Thomas Man died in September 1690 and, like a butterfly that fails to break free from its chrysalis, the insect project came to an abrupt halt.

In May 1691, John Ray wrote to Richard Waller, Secretary of the Royal Society, saying – somewhat mysteriously – that although he had seen Francis Willughby's insect materials, he had long since given them to Sir Thomas, 'who will I suppose in time take care to publish them'.[70] This suggests first, that Ray viewed his own *History of Insects* as an independent project, and second, that Willughby's entomology book, thanks to Man and Sir Thomas, was close to reaching a publishable form. Yet ten years later it was still unpublished, and for some unknown reason the manuscript was by then at the Royal Society. In 1704, Ray was able to borrow it and began to incorporate some of its contents into his own book on which he was now working in earnest.

In his late seventies, this was an extremely difficult time for Ray, who was suffering from incurable suppurating ulcers on his legs. In 1700 he had written to Hans Sloane, Secretary of the Royal Society, telling him that he had 'little or no absolute intermission of pain'. Three months later he wrote again: 'I am now much worse, being by the sharpness of my pains reduced to that weakness that I can scarcely stand alone, so that I lay aside all thoughts of the *History of Insects* and despair even of life itself.' It may be no coincidence that he felt the cause of his ulcers to be invisible insects whose nests lay in the tumors in his legs. It is a measure of his fortitude, or perhaps the absolute need for some kind of distraction, that he persevered so assiduously with the insect project. It is clear, too, that even as his body fell painfully apart, Ray's mind remained as sharply focused as ever.[71]

The objective of the *History of Insects* was the same as the *Ornithology* and the *History of Fishes*: species descriptions, identification and classification. There was an important difference from the previous volumes, however, and one that marks Francis Willughby's contribution to entomology as especially novel. This was his interest in insect generation – reproduction and development – and, in particular, the metamorphosis from larva to adult. The presence or absence of metamorphosis was the crucial criterion in Willughby's classificatory system, and Ray made it the foundation of the classification in *Historia Insectorum*.

After Ray died, in January 1705, with the text still unfinished, it was hoped that his friend Samuel Dale would complete the manuscript for publication, but for whatever reason he failed to do so. The manuscript was then given to another of Ray's friends, William Derham, who in turn did nothing either, eventually passing the papers to Hans Sloane at the Royal Society in 1708. Even there, the manuscript's fate was uncertain. Tancred Robinson, physician and naturalist and Fellow of the Royal Society, one of many who had played an important role in seeing the *History of Fishes* through to publication twenty years previously, commented to Martin Lister that Ray's papers 'lye in Dr Sloane's hands, and when they will see a resurrection I cannot divine'.[72] In 1709 a Royal Society committee took control, deciding that even though the manuscript was incomplete and devoid of any images, it should be printed. When they duly arranged for its publication, Derham, who knew of the manuscript's convoluted history, suggested to the committee that they should delay publication until they had consulted Sir Thomas Willoughby, but Derham was overruled. The *History of Insects* appeared in 1710 with Ray's name as the sole author, a decision, it seems, made by the Royal Society committee, rather than by Ray before he died.

Shortly after the book appeared in print, William Derham met Emma Child and her daughter Cassandra, who told him in no uncertain terms that they and Sir Thomas 'took it ill' that Francis Willughby's insect papers were published solely under Ray's name. Emma said also that they considered it 'surreptitious' that the book had been published without their knowledge, pointing out that Sir Thomas and Dr Man had spent nearly two years on the project and had prepared numerous plates 'wch are now lying by them'.[73] In an attempt to smooth things over, Derham suggested that a new edition might be produced that included Willughby's illustrations. It didn't happen and the prepared plates, like the original Willughby manuscript, subsequently disappeared. The hope now is that they may be lying in someone's attic or hidden away in a library and will eventually be rediscovered.

The lengthy introduction that also serves as a comprehensive list of contents to the *History of Insects* – what Ray entitled the Prolegomena – provides us not only with a sense of what's in store but also the scale of the task that Willughby had set himself. There are currently thought to be between one and three million species of insects in the world, of which only a third have been described; there are over 20,000 species of insect in Britain alone.[74] Even though many fewer were recognised in Willughby's day, the number was far in excess of the known number of birds or fishes. An additional difficulty was that, in contrast to birds and fishes, most insects (and indeed, most invertebrates) had no names. Willughby and Ray deal with this by using 'polynomials' or what I have called descriptors: short phrases such as 'A large long-horned Beetle, with big horns, jointed and bent back'; 'A water-beetle with ridged or channelled elytra'; 'An elegant Crane fly, with black back and elytra, a cross-shaped belly, with wings marked with a dark spot, light and gleaming'.

Insects had no names partly because people were – up to this point – much less interested in them than they were in birds. Birds were familiar and useful; they were caught and kept for their song, or killed and cooked, or taken as a cure. Because different bird species were used for different purposes they had been allocated names to distinguish them. Birds – and fish, for similar reasons – were therefore far ahead of insects in terms of possessing common names. Confusion persisted whenever the same bird was known by different names in different parts of the country, but overall, as the *Ornithology* makes clear, all British birds possessed a common name. In contrast, only a handful of species in the *History of Insects* had a common name.

The different ways in which Willughby and Ray treated birds, fish and insects in their three volumes reflect the way they viewed the animal (and plant kingdoms) as a whole. This isn't explicit in the *Ornithology*, *History of Fishes* or *History of Insects*, but Ray spells it out in his book *The Wisdom of God*, written during the late 1680s – perhaps before he even started on the insect work. He refers to a 'scale of perfection' that, in descending order, runs from beasts (quadrupeds) to birds, fishes, insects and plants. Beasts are more perfect than birds, which in turn are more perfect than fish, and so on.

Ray may have borrowed this idea of a scale of perfection from his friend John Wilkins, who wrote:

> It may be observed to be amongst these [i.e. fishes, birds and beasts] ... that the more perfect kinds are the least numerous. Upon which account, insects being the most minute and imperfect, and some of them (perhaps) of a spontaneous generation, are of the greatest variety, tho by reason of their littleness, the several species of them, have not hitherto been sufficiently enumerated or described, by those Authors who have particularly applyed themselves to this study.[75]

Although based on the medieval notion of 'the great chain of being' – a hierarchy of life forms decreed by God – Ray's interpretation of the scale of perfection is remarkably free of religious connotations and simply reflects the number of species, their size and their distinctiveness. There are more birds than there are beasts, more fish than birds, and more insects than fish. Based on his friend Martin Lister's reckoning, Ray presumed that there were at least 1,800 to 2,000 species of insects, which he says 'for numbers [may] vie even with Plants themselves'. By comparison, Willughby and Ray thought there might be around 500 species of birds.[76]

The scale of perfection also reflects size, with beasts being generally bigger than birds, and so on. Indeed, as Ray points out, many insects (meaning little animals, in general) are so small they can be seen only by using a microscope, as shown by 'Mr. *Lewenhoek* of *Delft* in *Holland* ... whose Observations were confirmed and improved by our Learned and Worthy Country-man Mr. *Robert Hook*.'[77] Finally, the less perfect the organism, the less distinct it is from other organisms of the same type.

The degree of 'perfection', the number of species and their size (in particular the small size of many insects) all conspired to make each level on the scale less tractable and less familiar than the one above it. Distinguishing different insect species was much more difficult than it was for fish, or birds or beasts, which in turn meant that it was also much more difficult to assign them common names. Even so, Willughby and Ray's use of 'descriptors' was an advance over previous efforts, including their own. When they produced a classification of insects for John Wilkins's *Essay* in the 1660s, they limited themselves to categories, such as 'dragonfly', without attempting to distinguish different species. The use of descriptors in the *History of Insects* was an important step in the great chain of scientific progress, and it was crucial in allowing Linnaeus to create his binomial system of scientific nomenclature in the mid-1700s.

Inspired by Willughby and Ray's work, entomology became an increasingly popular topic for investigation, and it was James Petiver who began the process of assigning common names to insects. Beginning with butterflies and moths, Petiver granted them names such as lappet, prominent, tussock, fritillary, argus and admiral (which later became red admiral) in his book *Papilionum Britanniae Icones*, published in 1718.[78]

The insect volume contains a tantalising section on 'The poison of Spiders' that is relevant to ornithology. Ray, who had suffered from an irrational fear of spiders since childhood, tells us that those in England are neither 'very injurious or dangerous, but they are not absolutely without poison'. He then cites an experiment conducted by 'our most distinguished Harvey [William Harvey]', who pierced his hand with a needle, and then after rubbing the point of the needle on the jaws of a spider, pricked himself again. Unable to feel any difference between the two points, he nevertheless noticed that the one with the spider venom 'was soon covered with redness, heat and inflammation, as if it were girding itself up for a fight and the storming of a harmful ill'. Ray goes on to speculate about how animals cope when they ingest venomous spiders, citing the work of the Italian physician Francis Redi, who stated that 'In the stomach however the juices with which it [the venom] is mixed lessen its power, & weaken it: and that mucus by which the lining of the stomach is covered within blocks its way so that it does not touch or injure it at all.' Ray then says something that seems to contradict my impression of how he and Willughby worked: 'For most small birds they [spiders] are a delicacy: for they seize all types of spiders indiscriminately and for the sake of observing this I have reared small birds in cages.' Earlier, I suggested that their lack of observations of cage birds had restricted Willughby and Ray's ornithological horizon, but here is Ray saying that he, at least, reared birds in cages in order to conduct experiments. Perhaps he did this while working on the insect book, long after Willughby's death. Anyway, he is correct that many insectivorous birds eat spiders, and several of them, including great tits, blue tits, and pied and red-breasted flycatchers, we now know selectively feed their five- to six-day-old nestlings on spiders, suggesting that at this particular stage of development their chicks require some specific nutrients best provided by spiders.[79]

The *History of Insects* appeared only in Latin. Ray's successors recognised it as a valuable contribution to entomology, and Linnaeus identified Ray and (with typical immodesty) himself as the best describers of insects. As Brian Ogilvie has said: 'Though much in the *Historia Insectorum* is Ray's, Willughby deserves recognition as one of the outstanding proto-entomological practitioners of the seventeenth century.'[80]

We don't know how many copies of the insect book were published, nor how the general public greeted it. I imagine its reception to have been cool, for even knowing of its historical significance today, with no illustrations the book seems relatively unattractive. There *could* have been illustrations had the Royal Society worked more closely with Sir Thomas Willoughby and used the images created by Thomas Man. But following its very long gestation the Society was desperate to see the book published and anxious

The only surviving drawings of insects prepared either by Francis Willughby's son, Thomas, or by Dr Thomas Man.

to avoid the kind of debt that the fish illustrations had incurred. A single set of tiny ink and wash images of bees and wasps, probably from Thomas Man, survives, escapees from the protracted and perilous journey endured by Willughby's other insect materials. Those images give a sense of what the book might have looked like: attractive, but not in the same league as the fishes, nor of the insect illustrations in Robert Hooke's *Micrographia* of 1665.

In the early 1800s William Kirby – known as the father of insect studies – appreciated the *History of Insects* and, recognising the important contributions made by both Willughby and Ray, cited them together with Martin Lister as 'stars of the first magnitude and the brightest lustre', diffusing 'new light over every department of natural history'. Kirby also acknowledged Willughby's perceptive observations on leaf-cutter bees, and took Linnaeus to task for not reading the *History of Insects* carefully enough with regard to bees. It was Kirby who called the leaf-cutter bee – whose larvae Willughby had reared – with punnish humour the 'willow bee' *Apis willughbiella*, thus commemorating Willughby's entomological eminence.[81]

The fact that biologists today are less familiar with either Willughby the ichthyologist or entomologist than they are with Willughby the ornithologist is not simply due to the diminished perfection of fish and insects, or the lack of English translations of those works. It is largely because the originality of Willughby and Ray's approach in the study of natural history lay in the *Ornithology*. The *Fishes* and *Insects* volumes contain much less first-hand material, and the information within those two books is often presented in a rather tedious, formulaic manner – certainly in the *Fishes*. As his biographer Charles Raven has said, Ray's heart wasn't really in the fish book: he did it out of a sense of duty to his friend.[82] For the *Fishes* book, too many cooks spoiled the bouillabaisse: there were simply too many authors and production managers to create a suitably digestible dish. And if Ray's heart wasn't in the fishes, it certainly wasn't in the insects. Worn out with age and the excruciating pain from his ulcerated legs, he ploughed on to the bitter end to ensure that he did justice to his friend's endeavours.

Birds had led the way, and it was inevitable therefore that the books on fishes and insects would seem less novel. They were both still a considerable achievement, full of valuable information, and each a landmark publication in the study of those two animal groups, but they lacked the iridescent sparkle of the *Ornithology*.

A Sustained Finale: Willughby's Buzzard Takes Flight

Francis Willughby would have been thrilled by the knowledge that his curiosity, industry and innovation were immortalised in the names of a fish, a bee and an entire genus of exotic plants. I doubt whether the absence of a bird from this list would have worried him unduly, for I am sure that he viewed his efforts across the animal and plant kingdoms as fairly even, even if we don't. He was, after all, interested in all aspects of the natural world.

The names Willughby's charr, Willughby's bee and the plant genus *Willughbeyi* were ascribed by later experts in those taxonomic groups. These were people interested in the history of their subject and with some certainty of knowledge regarding Francis Willughby's contributions, not only to that particular area of natural history, but to the entire field.

Some scientists, artists, writers and musicians gain their reputation while they are alive, and that reputation may subsequently grow, be sustained or, as in the case of the biologist Paul Kammerer or the art critic John Ruskin, decline. In other cases, reputations are made only in retrospect when the full extent and significance of someone's achievements are revealed.

Francis Willughby's reputation is unusual in the way it has fluctuated over time. Certainly, he was well respected by Royal Society contemporaries during his lifetime, but he was just one of several clever 'new philosophers'. His early death surprised and dismayed his colleagues, and forced them to make an assessment. They knew he was good, but apart from his few letters in the Society's *Transactions* there was rather little hard evidence of Willughby's talents. Instead, they had to defer judgement until John Ray had pulled together and published Willughby's works.

The *Ornithology*, appearing just four years after Francis's death, while memories of him were still strong, did much to consolidate his reputation, not least because of the eulogy that John Ray included in the Preface. No one knew Willughby like John Ray. His assessment of him was undoubtedly coloured by his sense of loss and by his commitment to the Willughby family, to whom he was dedicated, but also on whom he was dependent. Nonetheless, Ray's appraisal of his friend's personality and abilities was probably very honest. Certainly, no one during Ray's lifetime wrote anything to the contrary.

The three decades it took Ray to complete his friend's works both contributed to and detracted from Willughby's reputation. Ray's own stature resulting from his numerous published works such as the *Wisdom of God* was continuing to rise, while the public memory of Francis Willughby as the author of the *History of Fishes*, and of the *History of Insects* where he wasn't even identified as an author, was, in some quarters at least, beginning to wane.

The family took rather longer to appreciate Francis's achievements. Initially, this was because of Emma and Josiah Child's messy marriage and their even messier relationships with Francis's and Emma's children, Francis junior, Thomas and Cassandra. It was only after the children had grown up and freed themselves from Child's selfish and puerile ways that they began to understand and value what their father had accomplished during his brief life, and started to preserve what was left of his papers and collections.

᚛᚛

Throughout the eighteenth century, naturalists such as Carl Linnaeus, Thomas Pennant, the Comte de Buffon and Gilbert White all continued to applaud and cite Willughby with regard to both the *Ornithology* and the *Fishes*. Indeed, it is hard to find a natural history book from that period that doesn't mention him – and in glowing terms. Francis Willughby was *the* natural history authority. But, in 1788, after purchasing Linnaeus's collections and founding the London-based Linnaean Society, the botanist Sir James Edward Smith redirected the spotlight away from Willughby and onto John Ray. In his introductory address to the new Society entitled the 'Rise and Progress of Natural History', Smith stated:

> These illustrious friends [Willughby and Ray] labored together with uncommon ardour in the study of nature, and left scarcely any of her tribes unexplored. But death, which so often disappoints the fairest hopes, cut off the former in the prime of life, before he had digested the materials to the acquisition of which he had devoted his youth; and they might well have been lost to the world and his name have perished with them, but for the faithful friendship and truly scientific ardour of Ray.[1]

He continues:

> So close was the intercourse between these two naturalists, that it is not easy to assign each his due share of merit. Indeed, Ray has been so partial to the fame of his departed friend, and has cherished his memory with such affectionate care, that we are in danger of attributing too much to Mr Willoughby [sic], and too little to himself.[2]

In response to Smith's pronouncement, 'The ornithologists rose in wrath to denounce this impertinent botanist and defend their patron saint.'[3]

The first to come to Willughby's defence was William Swainson, whose father John had been one of the original members of the Linnaean Society. In 1834, some six years after Smith's death, Swainson, who had a reputation for being fractious, published a 460-page book entitled *A Preliminary Discourse on the Study of*

Natural History in what seems to me to be an attempt to usurp Smith's earlier account. In it, Swainson declares Willughby to be 'the most accomplished zoologist of this or any other country; for all the honour that has been given to Ray, so far as concerns systematic zoology, belongs exclusively to him. He alone is the author of that system which both Ray and Linnaeus took for their guide, which was not improved by the former or confessed by the latter.'[4]

To ram home his case, Swainson continues:

> It has been customary for writers to represent Willughby more as a wealthy and intelligent amateur than as an original thinker; as the disciple and pupil of Ray in zoological pursuits rather than his master and instructor ...
>
> But for the patronage and protection of Willughby, he [Ray] might probably have done little or nothing in science; and had he not been the editor of his patron's works, his name *as a zoologist*, would have been far inferior to that of Lister, for he had neither the talents of the first, or the originality of the last: yet he labored conjointly with both, and his name assumes a superiority from the variety of subjects he wrote upon, and from the number of works which bear his name, either as author or editor.[5]

Swainson's final flourish is to say that Ray 'cannot be said to have possessed great genius', a statement he moderates somewhat by adding that he had 'sound judgment, great zeal, unwearied application; – a pious, amiable and benevolent spirit, ever ready to acknowledge and praise the labours of others.'[6]

Two very ardent and uncritical disciples of Swainson were Neville and Charles Wood, precocious brothers from an eccentric, utopian socialist family. With misplaced aspirations to ornithological greatness, they each – at the tender age of eighteen – produced a book on birds, only months apart. And what curious books they are, not least because – out of undoubted competition – they hardly mention each other.[7]

And what a hodge-podge of ornithology! In a well-intentioned effort to make birds' common names more logical, they suggested some sweeping changes. They weren't alone: tinkering with the names of birds was rather like a parlour game. Perversely, however, what the Woods proposed was often anything but logical and rarely an improvement. Among other things, they suggested changing the name of the common blackbird to the 'garden ouzel' and the continental race of the yellow wagtail to the 'blue-headed oatear'.[8]

Bizarrely, both brothers' books included a critical overview of the ornithological literature. With what authority one wonders? At such a young age, however well tutored and educated, they simply didn't have the experience (even though today few would disagree with their evaluations). Their assessment of past bird books is horribly polarised, with authors classed either as useless or brilliant. And Willughby, of course, is brilliant.[9]

The Woods' decision to single out Willughby as the champion of bird studies was clearly based on their naive (and unreciprocated) hero-worship for William Swainson, who at that time was Britain's top bird man. Charles Wood wrote: 'Not only has Willughby been defrauded of his due with regard to the work [the *Ornithology*], but also the names given by him have been ascribed to more modern authors.'[10] And in a similar adversarial vein Neville commented, 'the amiable Ray, whatever he might be in botany, had very little merit as an ornithologist, the whole of his system … being the production of his friend Willughby; as is frankly acknowledged by Ray himself, and must therefore be true.'[11]

The onslaught was maintained by another eminent ornithologist, William Jardine, who in a long-winded and adulatory memoir wrote (after forty-three pages of preamble): 'It is now the place … to submit to the reader's attention the chief particulars in the history of the English Naturalist, Francis Willughby, Esq. of whom, although his name occurs in almost every treatise on Natural History, and often with high commendation, yet no *Memoir* has been published calculated to illustrate varied excellencies of his character, or to do justice to the genuine claims of his improvements in science.'[12] Jardine repeats and reinforces the views espoused by William Swainson and Charles and Neville Wood, as though all of them were independent authorities on Willughby.

More balanced was Alfred Newton in his history of ornithology – which appears as a mere 72,000-word Introduction to his *Dictionary of Birds* – a book that one reviewer described as 'the best book ever written on birds'. In it he states that Willughby, 'who at first the other's pupil, seems gradually to have become the master'.[13]

It is hardly surprising that there would be a backlash.

It came in the 1940s from Canon Charles Raven, theologian and master of Christ's College, Cambridge, in his wonderful and scholarly biography of John Ray. As a birdwatcher, botanist and entomologist guided by God, Raven, it seems to me, imagined himself almost as a reincarnation of Ray, determined to reassert and secure his reputation. Raven tells us that he 'collected nearly all the plants, birds and insects Ray records, and often in the same localities; and the thing that he tried to do, to reinterpret the faith of a Christian in the light of sound knowledge of nature, has been my continuous and chief concern.'[14]

Raven's densely packed book is a comprehensive account of Ray's life and work, but it is also designed to counter the view of Willughby as the hero. Raven is courteous but relentless in his defence of John Ray. He starts with four 'indisputable facts', which, briefly, are as follows: Ray's university record makes it clear he was a genius, 'in the Galtonian sense' (meaning, a natural genius?); Ray was responsible for collections made abroad, for observations on birds and fishes and the independent study of birds, fish and insects; he was 'scrupulously punctilious … in acknowledging … others'; and finally, he 'wrote with enthusiasm of all his friends', never uttering a word of criticism, adding: 'We could never gather from his letters that Lister was one of the most self-satisfied and obstinate of men, or that Sloane was a snob and a dilettante.'[15] These facts may indeed be indisputable, but they hardly inform the argument about the relative merits of Willughby and Ray since much the same could be said of Francis Willughby.

In what he seems to have set up as a kind of academic beauty contest, Raven then enumerates five points about Willughby. These are that: (i) he was 'a man of high principles, great fertility of mind and energy of disposition, keen, interested, enthusiastic' and in 'the language of the time a "virtuoso"'; (ii) he 'left very little original work'; (iii) he had a 'real gift for speculative ideas, which Ray did not always think sound or probable'; (iv) he 'contributed almost nothing to Ray's botanical work except in sharing in the experiments on the flow of sap'; and (v) his best work was on insects, which 'shows keenness, ability and a wide range of observations, but is elementary as compared with the best of Ray's work'.[16]

Throughout his book Raven never misses an opportunity to snipe at Willughby and is relentless in grinding down any opposition to Ray. At one point he says 'compared to Willughby, Ray has a clearer grasp of scientific method and a truer insight into the principles of natural history'.[17]

In his triumphant summing-up, Raven declares Willughby to have had 'less knowledge, patience and judgment' than Ray, and 'the evidence makes it certain that Ray was a scientist of genius and probable that Willughby was a brilliantly talented amateur'.[18] Touché. Job done: the ornithologist routed.

Charles Raven's elegant, wordy and adulatory account of Ray's life left readers in little doubt where the real talent lay in the joint Willughby-Ray enterprise: a view that remained unchallenged for over fifty years. Now, in the light of more recent research and new discoveries, we are better placed to assess Willughby's contribution. Although Raven was motivated by an enormous sense of injustice to Ray, he is correct in assuming Ray to be a genius, but too harsh in demoting Willughby to the rank of 'enthusiastic amateur'. In the strict sense of the word, Willughby *was* an amateur in that he had never held an academic position, as Ray had. That is why Willughby is sometimes described as a 'virtuoso'. We would probably make the same distinction today, albeit without using the term. But Raven's wording is inappropriate. His comment that

Willughby's entomological work was 'elementary' compared with the *best* (my italics) work of Ray is blatantly unfair because it does not allow for Willughby's early death. Raven is also wrong – as we have seen – to say that Willughby contributed almost nothing to Ray's botanical studies. Indeed, as we now know, it was Willughby who initiated the studies of the flow of sap.

We can excuse Raven, to some extent, on the grounds that he knew much less than we do today about the depth and breadth of Willughby's abilities, but in the broad context of the history of science this game of spot-the-genius is inappropriate and unhelpful. What is more obvious than ever now is that Willughby and Ray worked as a team. They had a synergistic effect on each other and it is unlikely that in terms of their zoological efforts either would have been as effective on their own. As it was, they transformed the study of zoology in general, and birds in particular. Theirs was the first true classification; they set a new standard in observation and description, and their encyclopaedia formed the foundation on which all subsequent ornithology was built.

In 1909 the Nobel Prize-winning chemist Wilhelm Ostwald published a perceptive account on the psychology of scientists, whose extreme types he calls 'romantics' and 'classics'. The romantic's peculiarity, he said, 'lies not so much in the perfection of each individual work as in the variety and striking originality of numerous works following one another in rapid succession, and in the direct and powerful influence he has upon his contemporaries'. The sure-fire way to identify the classic, on the other hand, he says, is 'the all-round perfection of each of his works … and … need to stand unblemished in the public eye'. Ostwald finishes his description by saying 'the speed of mental reaction is a decisive criterion for determining to which type a scientist belongs. Discoverers with rapid reactivity are romantics, those with slower reactions are classics.'[19]

This is remarkable, for it is almost as if Ostwald had the curricula vitae of Willughby and Ray in front of him: Willughby with his swift succession of interests in birds, fish, insects, seeds, sap, games, mathematics and linguistics; Ray with his perfectionist's eye for detail, and the tenacity and endurance to bring to fruition whatever they started together.

It is now agreed that there is no single way of doing science and that both of Ostwald's types are equally important in driving progress; and, of course, these are the extremes. Most scientists exhibit some of both traits and lie towards the centre of that particular continuum.[20] The secret of Willughby and Ray's success, however, resulted from their collaboration and the synergism borne out of their respective talents and contrasting personality types. In his eulogy to his friend in the *Ornithology*, Ray lists those features he identifies as Willughby's characteristic marks: 'quick apprehension, piercing wit and sound judgement ... and his great industry'.

ᏊᏊ

I have been living with Francis Willughby for almost a decade, trying to imagine – on the basis of frustratingly slim evidence – what it was like to pioneer a new way of viewing the natural world. Exhilarating for Willughby, but also for myself as I came to understand him better. The more I read and the more I talked about him with my colleagues, the more I liked Willughby. He was fragile too, but constantly straining at the leash, never still, aware of how much there was to do and how little time he had, regardless of his health. But the more I thought about Francis, the more I felt that – notwithstanding the adulatory eulogy in the *Ornithology* – there was something a bit unsettling about the way in which Ray presents Willughby to the world.

The *Ornithology*, we are told by Ray, and we know from much other evidence, was a joint effort, based on Willughby's notes and Ray's organisational and literary skills. But just consider – outside that effusive Preface – how often Ray says 'we' compared with how often he refers to himself: 'I'. If this was a true partnership and if Ray was genuinely trying to ensure his friend's reputation, surely he should have been using 'we' throughout, but he doesn't. Moreover, he sometimes contradicts his colleague; 'Willughby found this', but '*I* found this', he says. Consider also Ray's response to Willughby's novel questions about birds: Ray is dismissive, almost indifferent.

He certainly does not use the queries to bolster Willughby's reputation. He doesn't say, for example, 'look at these ingenious questions that might inform or inspire future generations of ornithologists'. Consider also their classifications: Ray refers to Willughby's as 'a method of his own contriving'. And finally, throughout the *Ornithology*, Ray grumbles about Willughby's unnecessarily detailed descriptions of plumage.

One explanation is that Ray is simply being honest: that was the ways things were. But it isn't how you'd expect someone to celebrate the life of a prematurely dead friend. Ray's feelings towards Willughby may have partly been because he was exhausted by the work he had to do to prepare the *Ornithology*, which he says was more than he would willingly have undertaken.

Another possibility is that Ray is merely setting out his own stall – he had, after all, his own future and livelihood to think about. He wanted to become an authority on the natural world.

These niggles are neither unreasonable nor unrealistic, for great friendships and marriages often survive and thrive despite minor irritations, like the bramble thorn that leaves its tiny tip embedded beneath the skin of one's hand.

History tends to celebrate those who make discoveries that change the world: the laws of gravity, natural selection, penicillin and the structure of DNA. Willughby and Ray changed the world, but without making any single major discovery. What they did do, and it was just as valuable, was to create through their schemes of classification and scholarship a way of studying the natural world.

One way to evaluate Willughby's ornithological success is to compare what he and Ray identified and described with what we know today. They did pretty well, identifying and describing around 90 per cent of the roughly 200 birds regularly (if not always commonly) encountered in England and Wales (where they travelled). They missed several warblers, the firecrest, the pied flycatcher and the bearded tit, and struggled to distinguish different gulls and waders. They also overlooked Montagu's harrier, Bewick's swan and the Slavonian grebe, while neither the storm petrel nor the fulmar (which in the seventeenth century was probably confined to St Kilda) are mentioned in the *Ornithology*. Two species that I was surprised they got were the snow bunting, referred to as 'the great pied mountain finch or lesser brambling', shot by Willughby in Lincolnshire one winter, and the secretive, visually non-descriptive grasshopper warbler: 'A titlark that sings like a grasshopper.'[21]

Perhaps not surprisingly, Willughby and Ray were rather less successful with continental birds, identifying only about half of the seventy additional species they could potentially have encountered on their travels. Notable omissions include the booted, Bonelli's and short-toed eagles and the black kite, several terns (notoriously difficult), middle-spotted and three-toed woodpeckers, the scops owl and the pin-tailed sandgrouse. Successes included the citril finch – the *verzellino* – discovered (as a cage bird) by Ray in Rome in 1664, and the spotted crake that Willughby found and described in a Valencia market while apart from Ray and his colleagues.

Willughby and Ray were probably the first to describe the common scoter, Manx shearwater, crested pochard, red grouse, spotted crake and honey-buzzard. Ray obtained a specimen of the common scoter – a difficult-to-observe duck that spends almost all its time on the sea – at Chester while he and Willughby were staying with their friend, Bishop Wilkins, in 1671. Despite their having received a scoter from Francis Jessop a few years previously, Ray claims this one for himself, stating in the *Ornithology*: 'I found the male of this kind at Chester, killed on the sea-coasts thereabouts, and bought in the market by my Lord Bishop Wilkins ... this bird hath not as yet been described by any author in extant print that we know of.'[22]

The Manx shearwater (or Manx puffin as it was then known), seen by Willughby during their visit to the Calf of Man in 1662, barely counts as a discovery because the bird that Francis described was a chick – little more than a ball of fluff. It was only later that Ray saw dried specimens of the adult bird at the Royal Society's repository and in John Tradescant's famous cabinet at Lambeth, when he was able to describe it properly – but even then he failed to recognise it as the same species that Thomas Browne had maintained in captivity.

The great red-headed duck – the red-crested pochard – was Ray's discovery entirely: he found it in the market in Rome when Willughby was elsewhere in Italy: 'I never happened to see it elsewhere, neither do I find any description of it, or so much as any mention made of it in any book. Where it lives and breeds I know not.'[23]

Willughby's spotted crake, also – sadly – doesn't count because there is no firm evidence that he recognised it as distinct: that is, specifically distinct from the bird later known as Baillon's crake, which he and Ray had seen and described in Italy.[24]

When it comes to the honey-buzzard we are on firmer ground. We do not know for sure, but I suspect it was Willughby who made the discovery. It was his obsession with detail that would have enabled him to see that the bird he had in front of him was new and had not previously been described.

There are many groups of bird species whose outward appearance makes them difficult to distinguish: certain warblers, harriers, wading birds (especially outside the breeding season), pipits, marsh and willow tits, the Eurasian tree creeper and short-toed tree creeper, and so on. Raptors were especially difficult because they exhibit a huge amount of individual variation in their plumage colour.[25] The common buzzard, for example, varies from being almost completely white to pure chocolate brown, with everything in between. In this respect it is all the more remarkable that Willughby should notice a different sort of buzzard. The published account, obviously written by Ray, finishes in a characteristically modest and understated way: 'It hath not as yet (that we know of) been described by any Writer.' They knew it was new.[26]

All of Willughby's diagnostic characters of the honey-buzzard are absolutely correct, but he seems to have overlooked (or Ray overlooked when writing up the account) what is now considered one of the most interesting, if not the most obvious, features of that species: the curious, scale-like feathers on the face. The honey-buzzard's principal prey is the larvae of social wasps, which it finds by following the flight-lines of adult wasps as they return to their nest. The bird then alights on the ground and digs out the nest using its short strong legs and claws. Adult wasps, as everyone knows, are ferocious in the defence of their nest, but the honey-buzzard's distinctive face-feathers have evolved to protect it from the stings. Willughby and Ray knew that the honey-buzzard preys on wasps' nests for they found fragments of them in a honey-buzzard nest they examined.

᛭ ᛭ ᛭

There's something rather comforting about the honey-buzzard's name, conjuring up, as it does, images of a bird with warm and mellow plumage, or of one that instead of feasting on live prey, ingests the pure and natural sweetness of the bees' selfless labours. But this is wrong, for the honey-buzzard neither seeks nor eats honey in the wild.

It seems likely that the name was one given by local people, and Willughby and Ray intimate as much in the *Ornithology*, based on the erroneous but widespread assumption that wasps, like bees, create and store honey.

The idea that Willughby had discovered the honey-buzzard was later challenged by Alfred Newton, the pompous, blustering, misogynistic doyen of Victorian ornithology. His fame rests on his *Dictionary*, helping to found the British Ornithologists' Union, encouraging young ornithologists and being an undoubted scholar. He accumulated an impressive library (still extant) and he knew his books inside out. It is a measure of the man that he annotated many of his books, correcting the author's grammar, punctuation or statements of fact. He considered himself the expert on all things ornithological.

Under the heading 'honey-buzzard' in his *Dictionary of Birds*, Newton says that this was a bird Willughby 'thought he was describing for the first time; but herein he was wrong, for it was the *Boudree* of Belon (1555)'.[27]

I went to check Pierre Belon's encyclopaedia, published over a century before the *Ornithology*, but was confused, for it was far from obvious that the bird Belon called the *Boudree* was a honey-buzzard. I consulted colleagues who were better able than I to read and interpret Belon's sixteenth-century French text, and after much hesitation, deviation and repetition by Belon, it became clear that while he must have seen honey-buzzards, the description in his book is actually a muddled mix of honey-buzzard and common buzzard features. Newton's claim was not valid.[28]

The honey-buzzard's current official name is the 'European honey-buzzard' and was adopted, presumably, to distinguish it from the 'Oriental honey-buzzard' – a separate, closely related species that occurs in eastern Asia.

But 'European honey-buzzard' may be the most inappropriately named of all birds. As already noted, it doesn't eat honey. 'European' is wrong, too, for the species breeds across Eurasia and winters in Africa. It is just possible that 'buzzard' may also be inappropriate, since recent molecular studies suggest that rather than being a close relative of the common buzzard, as is widely assumed from its appearance, it may be more closely related to some of the tropical kites.

Time for a change then. Time to abandon 'honey' and 'European', but perhaps not quite time to shrug off the 'buzzard' – the molecular studies are still struggling for clarity here. I think we should rename this species 'Willughby's buzzard'. It is, after all, rather odd that Willughby, who is best known as an ornithologist, should have a bee and a fish named after him – but no bird.

Changing the common names of birds is not something one should do lightly, but in this case it seems entirely justified. The present name is wrong, and Willughby made an outstanding contribution to ornithology. He deserves to be remembered by a link to the bird he discovered.[29]

Let us celebrate Willughby's life and achievements by according him this additional twenty-first-century accolade: Willughby's buzzard.

Acknowledgements

First and foremost, I wish to thank the Middleton family: the late Lord and Lady Middleton, their son, Michael Willoughby, 13th Baron Middleton, and his son and daughter-in-law, James and Cara Willoughby, for their generosity, help and encouragement throughout the Willughby project.

Writing is a solitary occupation, but the research one does before that can be wonderfully sociable. The Willughby Project, funded by an International Network Grant from the Leverhulme Trust, to whom I'm extremely grateful, proved to be just that. This volume would not have been possible, nor half so much fun, without the extraordinary contributions made by the members of the Network: Isabelle Charmantier, David Cram, Meghan Doherty, Mark Greengrass, Daisy Hildyard, Dorothy Johnston, Sachiko Kusukawa, Brian Ogilvie, William Poole, Christopher Preston, Anna-Marie Roos, Richard Serjeantson, Paul Smith and Benjamin Wardhaugh. Our two meetings, deep in the heart of rural Derbyshire in April 2012 and March 2014 respectively, were among the most exhilarating of my academic career. I am grateful to my Willughby colleagues for helping to create the most extraordinarily stimulating project and I feel immensely privileged to have worked with them. I thank them all for answering my questions, but Isabelle Charmantier, David Cram, Mark Greengrass, Dorothy Johnston and Anna-Marie Roos deserve special mention for all the help they so willingly offered with the present volume.

I am also indebted to the following friends and colleagues who provided assistance and advice: Jane Amat, Sue Barnes,

Jonathan Barry, Rob Bijlsma, Patricia Brekke, Tony Campbell, Abigail Cobley, Mark Cocker, Nigel Collar, Michael Collins, Fred Cooke, Mark Dorrington, Tom Finch, Nathan Flis, Brian Follett, Gareth Fraser, Nick Fry, Emily Glendenning, David Horobin, Jim Horsfall, Ragnar Kinzelbach, Alexander Lee, Louise McCrickard, Peter Marren, Chris Mattison, Matt Merritt, Keith Moore, Mats Olsson, Philip Oswald, Robert Pearce, Rebecca Philips, Pietro Pizzari, Tom Pizzari, Michael Quinn, Roger Riddington, Douglas Russell (at the Natural History Museum, Tring), Karl Schulze-Hagen, Jon Slate, Helen Smith, Claire Spottiswoode, Tilli Tansey, Gavin Thomas, Jamie Thompson, Carlo Violani, John Wade, and the staff in the Manuscripts Department and their colleagues in IT services at Nottingham University Library.

The following very kindly located or provided images: Mark Bentley, Brown University, Simon Butler, Hayley Cotterill, Linda DaVolls, Ian Dillamore, Mark Dorington, Jeremy Early, Corrine Fawcett, Tom Finch, Jean Goodwin, Andreas Hartl, Klaus Nigge, Linda Shaw, Ann Sylph (Zoological Society of London), Katrina van Grouw, and Ray Wilson. I am especially grateful to Lord Middleton and his family and the University of Nottingham (Manuscripts and Special Collections) for permission to reproduce images from the Middleton Collection and from their home.

My friends Dorothy Johnston, Mark Greengrass, Bob Montgomerie and Jeremy Mynott each read the entire manuscript and I thank them for their critical and constructive comments, without which the book would have been much poorer.

I am extraordinarily privileged to have Felicity Bryan as my agent; her unfailing support and enthusiastic advice have been inspirational. I thank Michael Fishwick, my editor, and his superb team at Bloomsbury, and last but by no means least, I thank Richard Mason for his excellent copy-editing.

Finally, I started writing this volume in the mountains of Andalusia close to where Francis Willughby passed by in 1664. Willughby heartily disliked Spain and told his friend John Ray not to bother visiting (unless he had 'a mind to an Andalusian whore' – a joke). In contrast, I love Spain and for the last twenty-five years its people, vistas and wonderful wildlife have inspired my writing.

Appendix 1

A Timeline of Francis Willughby's Life

Date	Willughby Event	National Event
1635 22 November	Francis Willughby (FW) born at Middleton.	**1618–48** Thirty Years War (Europe); **1642** start of English Civil War; **1646–8** Witch fever in East Anglia: dozens of women killed; **1649** Charles I beheaded; **1651** Thomas Hobbes publishes *Leviathan*; **1652** Tea arrives in England.
1652	Enters Trinity College, Cambridge.	**1654** James Ussher, Archbishop of Armagh, proposes 4004 BC as the date of creation.
1656–9	Graduates with BA (1656). Graduates with MA (1659).	**1656** Burning alive for murder abolished in Britain, but burning for adultery still legal; **1657** Accademia del Cimento formed in Florence; **1658** Cromwell dies.

Date	Willughby Event	National Event
1660	John Ray (JR) enlists FW's help with Cambridge plant survey. FW visits Oxford University libraries to read natural history volumes.	Restoration of the Monarchy; Formation of Royal Society (RS).
1660 August	FW accompanies JR on trip through Yorkshire, Cumberland and Lancashire.	
1661 April–June		Charles II crowned king of England on 23 April; in June marries the Portuguese princess, Catherine of Braganza.
1661 21 November	At Royal Society, John Wilkins reads FW's letter on insect life histories; Wilkins proposes FW for membership.	
1661 December	FW admitted to Royal Society.	
1662 8 May–16 June	With JR and Philip Skippon (PS) on journey from Cambridge to Wales; on return FW leaves the others at Gloucester, becomes ill and returns alone. This is the journey on which the decision to overhaul the whole of natural history is made.	
1662 August	Act of Uniformity ends JR's Cambridge career.	
1662 October	FW at three RS meetings.	
1663 18 April	Start of continental journey with JR, PS and Nathaniel Bacon (NB), and two servants.	
1663–4	Continental journey.	
1664 mid-August–late November	FW travels through Spain.	Re-establishment of Anglican Church; Isaac Newton publishes on gravity.
1664 December	Returns to Middleton.	

Date	Willughby Event	National Event
1665 4 January	Presents astronomical observations at RS.	
1665 March	Conducts sap experiments.	Elizabeth Gaunt burnt at stake for treason.
1665 May–June	FW at RS: observes experiment on dog; discusses spontaneous generation and mites.	Great Plague of London just taking hold.
1665 December	Death of FW's father.	
1666 June–July	FW at various RS meetings.	September: Great Fire of London; Christopher Wren and Robert Hooke start redesigning London.
1666 20 October	Wilkins requests help with classification.	Merret publishes *Pinax*.
1666–7 winter	FW at Middleton with JR organising specimens.	
1667 28 March	Wilkins presents RS repository 'a kind of silken substance' taken from shellfish (?mollusc); FW 'affirmed that he had taken it himself out of a living shellfish, called pinna marina'.	May: Margaret Cavendish is first woman to be invited to a Royal Society meeting (once).
1667 27 June	FW at RS talks about coal.	Jean-Baptise Denys, physician to Louis XIV, performs the first human blood transfusion (sheep's blood) on a fifteen-year-old boy, who later dies; First edition of Milton's *Paradise Lost*; Peace of Breda, ending the Second Anglo-Dutch War.
1667 25 June– 13 September	Travels with JR to the West Country collecting specimens.	

Date	Willughby Event	National Event
1668 9 January	FW marries Emma Barnard.	England, Sweden and the Netherlands form the triple alliance against the French; Charles II gives Bombay to East the India Company; Francesco Redi publishes *Esperienze Intorno alla Generazione degli Insetti* (*Experiments on the Generation of Insects*); John Wilkins's *Essay towards Real Character* published.
1668 13 September	First son, Francis, born.	
1668–9 September–March	JR a regular visitor at Middleton: continues sap experiments.	
1669 April	FW and JR visit Bishop Wilkins at Chester. FW witnesses JR's porpoise dissection. FW taken ill.	Mount Etna erupts killing 20,000 people.
1669 29 May	Reports to RS on sap experiments with JR: read at RS on 10 June.	
1669 June–July	Writes to RS several times on sap.	
1670 20 January	Oldenburg (RS) writes to FW about insects wrapped in leaves.	Charles II and Louis XIV sign a secret anti-Dutch treaty.
1670 29 January	Replies to Henry Oldenburg regarding insects in leaves; sends some insect specimens; refers to expected collaboration with Wilkins on 'noble theory of motion', reported to RS on 10 February. This date: 'Mr Hooke being absent, read some letters, including some by FW.'	

Date	Willughby Event	National Event
1670 12 March	Writes to Oldenburg about sap and requests equipment (thermometers and barometers).	
1670 19 March	Oldenburg replies; includes mention of Tonge's curious observations on spiders.	
1670 23 April	Daughter, Cassandra, born.	
1670 28 April	At Wollaton with JR.	
1670 5 May	RS decides to send FW specimen of 'worm' wrapped in leaves (sent on 17 May).	
1670 19 July	At Middleton, writes to Oldenburg about Martin Lister's sap studies.	
1670 19 August	At Astrop with relatives to investigate bees in leaves; writes to Oldenburg; letter read at RS on 27 October.	
1670 2 September	At Middleton, writes to Oldenburg again about bees; letter read at RS on 27 October.	
1670 22 December	Lister writes to JR about FW's illness and recovery.	
1671 13 January	FW and JR write to Oldenburg from Middleton on bees and ants; letters read at RS on 19 January.	Louis XIV and the German Emperor sign a secret anti-Dutch treaty; Failed attempt by a drunk to steal the Crown Jewels in London.
1671 10 February	Sir William Willoughby dies; the will results in litigation.	
1671 spring	JR at Middleton ill with jaundice.	
1671 March	FW at Middleton writes to Oldenburg about sap; letter read at RS on 27 April.	

Date	Willughby Event	National Event
1671 13 May	Oldenburg writes to FW and sends insect specimen.	
1671 10 July	FW at Middleton writes to Oldenburg about hatching bees collected previously at Astrop.	
1671 24 August	FW at Middleton writes to Oldenburg about observations made with JR on insect parasitoids.	
1671 September	Visits Lister in York.	
1671 18 November	FW and JR in London dealing with litigation over Willoughby will.	
1671 late	FW abandons plan to visit North America due to ill health.	
1672 9 April	Second son, Thomas, born.	Charles II's Declaration of Indulgence; England and France declare war on the Dutch; John Bunyan released after twelve years in prison.
1672 8 May	At RS meeting in London, discusses parasitic worms.	
1672 3 June	Seriously ill.	
1672 24 June	Makes will.	
1672 3 July	Dies and is buried in Middleton church.	
1676	Latin *Ornithologiae Libri Tres* published.	Extremely hot summer in England; Antoni van Leeuwenhoek discovers the microscopic world.
1678	English *Ornithology* published.	Publication of the first part of Bunyan's *Pilgrim's Progress*.

Date	Willughby Event	National Event
1686	*Historia Piscium* published.	James II appoints four Catholics to the Privy Council; Around this time Robert Plot publishes *Natural History of Staffordshire*.
1705 17 January	JR dies.	Peter Courthope (FW's cousin) dies; Isaac Newton knighted by Queen Anne.
1710	*Historia Insectorum* published.	Following a hard winter, food is short in England.

Appendix 2

Brief Biographies of the Principal Players in Willughby's Life

FW = Francis Willughby, JR = John Ray, RS = Royal Society

Aldrovandi, Ulisse (1552–1605), Italian naturalist, author of *Ornithologiae* (1599–1603).

Antrobus, George (d.1708), curate for the Willughby family at Middleton (1665–76) and Wollaton (1679–1708).

Aristotle (384–322 BC), Greek philosopher, wrote extensively on natural history.

Aston, Francis, FRS (1645–1715), Secretary of the RS.

Bacon, Francis (1561–1626), precocious polymath; philosopher, statesman, author.

Bacon, Nathaniel (1647–76), pupil of JR, accompanied FW on continental tour; later colonist of Virginia.

Baldner (Baltner), Leonard (1612–94), 'fisherman', magistrate and amateur natural historian on the Rhine; produced (multiple copies of) an illustrated book on this topic.

Barnard, Emma (1644–1725), wife of FW; married 9 January 1668; second marriage to Josiah Child, but chose to be buried alongside Francis.

Barnard, Henry (1618–80), father of Emma, FW's wife.

Barrow, Isaac, FRS (1630–77), mathematician, developed calculus, Cambridge friend of FW.

Belon, Pierre (1517–64), naturalist, traveller, author of *L'Histoire de la nature des oyseaux* (1555); murdered in the Bois de Boulogne.

Bloot, Hugo de (1534–1608), Dutch librarian and historian.

Boyle, Robert, FRS (1627–91), natural philosopher and chemist.

Browne, Sir Thomas (1605–82), polymath, naturalist, author of *Pseudodoxia Epidemica*; his daughter painted images of birds borrowed by JR for *Ornithology.*

Charleton, Walter, FRS (1619–1707), physician and natural historian; published four images of birds including the pin-tailed sandgrouse.

Child, Sir Josiah (1630–99), merchant, politician, governor of East India Company; Emma's second husband after FW died.

Colonna, Fabio (1567–1640), Italian philologist and antiquarian.

Concublet, Andrea (1648–75), Marquis of Arena and from 1664 to 1668 organiser of the Academy of Investiganti in Naples.

Copsi, Fernandino (1606–86), Italian owner of a museum, whose collection was merged with Aldrovandi's in Bologna.

Courthope, Peter (1635–1705), cousin and friend of FW.

Crew, Sir Thomas (1624–97), politician, obtained image of pin-tailed sandgrouse now in the Middleton Collection.

Dale, Samuel (1659–1739), physician and botanist, friend of JR.

Dal Pozzo, Cassiano (1588–1657), prominent figure in Rome's intellectual circle; member of the Accademia dei Lincei; amassed the 'paper museum'.

Debes, Lucas Jacobsen (1623–75), Danish priest and writer; wrote on Faeroese seabirds.

Derham, William, FRS (1657–1735), clergyman, natural philosopher, author of *Physico-Theology* (1713); edited JR's correspondence.

Descartes, René (1596–1650), French philosopher, mathematician and scientist.

Dixie, Sir Beaumont (1629–92), husband of Mary Willoughby, sister of Sir William Willoughby of Selston.

Duport, James (1606–79), classical scholar, Trinity tutor of FW and others.

Ent, George, FRS (1604–89), anatomist, scientist and defender and promoter of William Harvey's works.

Fabricius ab Aquapendente, Hieronymus (1537–1619), naturalist, embryologist in Padua.

Faithorne, William (1616–91), artist and engraver; created the crayon portrait of JR and engraved some of the birds for the *Ornithology*.

Fallopio, Gabriele (1523–62), Italian anatomist; studied the head and reproductive system.

Gessner, Conrad (1516–65), Swiss naturalist and scholar; author of *Historia Animalium* (1551–8).

Hartlib, Samuel (1600–62), German-British polymath.

Harvey, William (1578–1657), physician to James I and later Charles I; discovered circulation of the blood; author of *Exercitationes de generatione animalium* (1651).

Hewitt (Hewett), Dr Anthony (1603–84), FW's physician in the weeks before his death.

Hill, William (1618–67), classics scholar, FW's teacher at the free school at Sutton Coldfield.

Hooke, Robert, FRS (1635–1703), curator of experiments at the RS; author of *Micrographia* (1665).

Howell, James (1594–1666), travel writer and historian.

Hulse, Edward (1631–1711), famous physician; friend of JR.

Huygens, Christiaan, FRS (1629–95), Dutch mathematician, interested in probability.

Jardine, William (1800–74), ornithologist.

Jessop, Francis (1638–91), naturalist, mathematician, based in Sheffield.

Jonston, Jan (1603–75), Polish scholar and physician, author of *De Avibus* (1657).

Kirby, William, FRS (1759–1850), entomologist and parson-naturalist.

Kircher, Athanasius (1602–80), self-promoting Jesuit polymath and prolific writer; dissected nightingales.

Leeuwenhoek, Antoni van (1632–1723), draper, tradesman, microscopist.

Lessius, Leonard (1554–1623), Flemish Jesuit author of *The Providence of God* (1631 in English).

Linnaeus, Carl (1707–78), Swedish botanist, zoologist and physician, responsible for the binomial system of naming, author of *Systema Naturae*.

Lipsius, Justin (1547–1606), Flemish, philologist and travel writer.

Lister, Martin, FRS (1639–1712), physician, natural historian and prolific author.

Man, Thomas (d.1690), educated at Jesus College, Cambridge, trained as physician at Utrecht; worked with Sir Thomas Willoughby on his father Francis Willughby's insect project.

Marcgraf, Georg (1610–44), German astronomer and naturalist, author of *Historia Naturalis Brasiliae*.

Marchetti, Pietro (1589–1673), surgeon in Padua.

Martyn, John (d.1680), London publisher; sole publisher for the RS.

Muffet, Thomas (1553–1604), merchant, natural historian, author of *Health's Improvement*.

Montalbanus, Ovidius (1601–71), curator of Aldrovandi's museum in Bologna.

Newton, Alfred, FRS (1829–1907), professor of comparative anatomy at Cambridge University; zoologist and ornithologist, author of *Dictionary of Birds* (1896).

Newton, Isaac, FRS (1642–1726/7), physicist, mathematician; among the most influential of all scientists.

Nidd, John (d.1659), colleague of FW and JR at Trinity College, Cambridge.

Oakley, Margaret (n.d.), married JR in 1673.

Okely, Mr (n.d.), possible relative of Margaret Oakley (above).

Oldenburg, Henry, FRS (c.1619–77), German natural philosopher, theologian, secretary of the RS and founding editor of *Philsophical Transactions*.

Olina, Giovanni Pietro (1585–1645), author of *Uccelliera* (1622).

Pepys, Samuel, FRS (1633–1703), Member of Parliament, naval administrator, diarist. President of the Royal Society (December 1684 to November 1686).

Petiver, James, FRS (c.1665–1718), apothecary, botanist, entomologist.

Pigafetta, Antonio (1491–1534), one of the 18 of 260 that set off around the world with Magellan in 1519 and survived the journey.

Piso, Willem (1611–78), Dutch physician and natural historian, co-author of *Historia Naturalis Brasiliae.*

Plot, Robert, FRS (1640–96), professor of chemistry at Oxford University, naturalist author of natural histories of Staffordshire (1686) and Oxfordshire (1705).

Raven, Charles (1885–1964), theologian, Master of Christ's College Cambridge, JR's biographer.

Ray, John, FRS (1627–1705), naturalist; mentor and friend of FW, and co-author.

Redi, Francesco (1626–97), Italian naturalist; father of parasitology and experimental biology.

Robinson, Sir Tancred, FRS (c.1657–1748), physician and naturalist; physician to George I.

Rondelet, Guillaume (1507–66), professor of medicine at Montpellier.

Scaliger, Joseph Justus (1540–1609), French philologist and historian; settled in Leiden in 1593 and worked at the university there until his death.

Skippon, Philip, FRS (1641–91), accompanied FW on continental tour; naturalist, Member of Parliament.

Smith, Sir James Edward, FRS (1759–1828), botanist, founder of Linnean Society of London.

Soest, Gerard (c.1600–81), portrait painter; painted both FW and his mother.

Swainson, William, FRS (1789–1855), ornithologist, entomologist and artist.

Swammerdam, Jan (1637–80), Dutch entomologist, microscopist; studied parasitoids of white butterfly caterpillars in the 1670s.

Sydenham, Thomas (1624–89), physician and pioneer of bedside medicine; later (1680) fell out with FW and JR's friend, the physician Martin Lister.

Theophrastus (c.371–c.287 BC), Greek, successor to Aristotle; prolific author, including on plants.

Todi, Valle da Antonio (n.d.), author of *Canto de gl'Augelli*.

Tonge, Israel (Ezerel) (1621–80), clergyman, correspondent of FW.

Turner, William (c.1509–68), physician and natural historian, author of *Avium praecipuarum* (1544).

Venette, Nicola (1633–98), French physician, author of *Traité du rossignol* (1697).

Walther, Johann Jakob (1604–79), superb bird artist, father of two sons (below).

Walther, Johann Georg (1634–97), illustrator of Leonard Baldner's book.

Walther, Johann Friedrich (n.d.), brother of Johann Georg Walther.

Wendy, Lettice (1627–96), FW's sister.

Wendy, Sir Thomas (1614–73), politician, husband of Lettice.

Wilkins, John, FRS (1614–72), clergyman, naturalist, Founder Fellow of the RS; Bishop of Chester.

Willoughby, Cassandra (1670–1735), FW's daughter; Duchess of Chandos: Cassandra Brydges.

Willoughby, Cassandra (née Ridgway) (1600–75), FW's mother.

Willoughby, Francis (junior) (1668–88), FW's first son.

Willoughby, Francis 'the builder' [(1546–96), FW's great-grandfather.

Willoughby, Katherine (1630–94), FW's sister.

Willoughby, Sir Francis (1588–1665), Francis the naturalist's father.

Willoughby, Sir Thomas, FRS (1672–1729), FW's second son.

Willoughby, Francis, FRS (1635–72), our Francis.

Zwinger (1533–88), Swiss physician and humanist, author of *Methodus Apodemica*.

Appendix 3

The Story of Willughby's Charr

Charr are salmon-like fish with a circumpolar distribution that live in cold-water lakes. Isolated from other populations since the end of the Ice Age, the charr in different lakes have evolved into almost separate species. One would like to say 'distinct' species, but therein lies the problem: they are not always distinct.

Willughby discovered the charr in the summer of 1660. In truth 'discovery' is too strong a word. It was more of an encounter, for the distinctive nature of the Lake Windemere charr was already well known to locals. John Leland, writing in the 1530s to 1540s had commented on 'a straung fish cawlled a chare' in 'Wynermerewath'.[1]

On hearing of the fish, Willughby and Ray quizzed local fishermen for information and obtained specimens to describe and dissect for themselves. What they gleaned was that Lake Windermere held two types of charr: 'The greate having a red belly they call the red Charre, and the lesser having a white belly which they call the Gilt or Gelt Charre.'[2] The fisherman knew that the red charr spawns in early winter, and the white charr in early spring, but still regarded them as two forms of the same fish; yet Willughby and Ray considered them separate species. At the right times of year Willughby and Ray could have cut some open, examined their gonads and verified that, of the red charr, those with a red belly were male and those with a mauve belly were female. As it was,

Willughby's diligent search for distinctive features was rewarded with the discovery that the red charr lacked teeth on its palate, an observation whose full significance became apparent only later.

Given that the red charr of Lake Windermere is now regarded as the most variable and diverse of all vertebrate species – that is, of *all* fish, amphibians, reptiles, birds and mammals – it is remarkable just how much accurate information Willughby and Ray were able to assemble.

When, two years after their Lake District trip, Willughby and Ray travelled to North Wales and encountered the torgoch, also a charr, they considered it identical to the red charr of Windermere. Later still, when they came across the 'carpioni' in Lake Garda, they felt that this fish was probably also the same as the white charr of Lake Windermere.

It was probably at Lake Garda that Willughby realised he had hit upon a feature that unambiguously distinguished the two types. Lake Garda's carpioni lacked the palatal teeth, just as Willughby had found for the red charr. It later turned out that not only do these little teeth on the roof of the fish's mouth distinguish different types of charr, they are also what separate salmon from trout. The white charr, or gilt as it is known, of Lake Windermere is effectively a trout, as is the carpioni of Lake Garda. Confused? Don't be (although professional taxonomists continue to be): long ago both salmon and trout became trapped in cold-water lakes, and during aeons of icy isolation evolved into charr. Or at least the salmon did; the white 'charr' is now considered a type of trout.

Long after Willughby and Ray were dead, the naturalist Thomas Pennant – best known as one of Gilbert White's correspondents – claimed the red and white charr of Lake Windermere to be separate species based on their distinct breeding seasons:

> This remarkable circumstance of the different seasons of spawning fish [charr], apparently the same ... puzzles us greatly, and makes us wish that the curious, who border that lake, would pay farther attention to the natural history of these fishes, and favor us with some further lights on the subject.[3]

The 1950s champion of charr research, Winifred Frost, applauded Pennant's novel use of the breeding seasons to separate the two Windermere charr, for at that time and for centuries thereafter most taxonomic distinctions were anatomical. But it was Willughby who pioneered the use of life-history traits, such as the metamorphosis in insects, or timing of breeding, as ways of distinguishing different taxonomic groups. There's little doubt in my mind that had Willughby lived to write his own account of the charr of Windermere (and elsewhere), he would have used both his anatomical observations (the presence or absence of palatal teeth) *and* the difference in breeding season to create a case for the red and white charr being distinct.

As it was, I imagine Willughby and Ray making notes during the Lake District trip, and filing them away – along with much else – for later inclusion in their planned *History of Animals*. After Willughby died and Ray had published *Historia Piscium*, under Willughby's name, Albert Günther, president of the Royal Society in 1875–6, composed a comprehensive overview of charr taxonomy and decided to honour Willughby with the 'discovery' of the red charr, naming it *Salvelinus willoughbii*:

> Willoughby [sic] is the first who with the practiced eye of an ichythiologist examined the charrs of England and Wales, devoting a separate article to their description.[4]

Appendix 4

Identification of Birds Listed on Page 61 and Named in the *Ornithology*

This includes the name listed by Willughby and Ray, the common name, scientific name and page number in the *Ornithology* (current names from F. Gill and D. Donsker (eds), 2016. IOC World Bird List (vol. 6.3), doi: 10.14344/IOC.ML.6.3.)

Local name	Common name, Scientific name (page number in the Ornithology)
Ars-foot	Great-crested Grebe *Podiceps cristatus* (339)
Bald buzzard	Osprey *Pandion haliaetus* (69)
Bastard Plover	Lapwing (Green Plover) *Vanellus vanellus* (307)
Black-cap	Black-headed Gull *Chroicocephalus ridibundus* (347)
Bohemian Chatterer	Waxwing *Bombycilla garrulus* (133)
Cock of the Mountain	Capercaillie *Tetrao urogallus* (172)
Common Grosbeak	Hawfinch *Coccothraustes coccothraustes* (244)
Coulterneb	Atlantic Puffin *Fratercula arctica* (325)
Daker-hen	Corncrake *Crex crex* (170)
Didapper	Dabchick *Tachybaptus ruficollis* (340)
Dun-diver	Red-breasted Merganser *Mergus serrator* (335)
Fern-owl	European Nightjar *Caprimulgus europaeus* (107)

Local name	Common name, Scientific name (page number in the Ornithology)
Flusher	Red-backed Shrike *Lanius curio* (88)
Gid	Jacksnipe *Lymnocryptes minimus* (291)
Glead	Red Kite *Milvus milvus* (74)
Gorcock	Red Grouse *Lagopus lagopus scotica* (177)
Green Plover	Golden Plover *Pluvialis apricaria* (308)
Greenland Dove	Black Guillemot *Cepphus grylle* (326)
More-buzzard	Marsh Harrier *Circus aeruginosus* (75)
Ox-eye	Great Tit *Parus major* (240)
Pool Snipe	Redshank *Tringa totanus* (299)
Puffin of the Isle of Man	Manx Shearwater *Puffinus puffinus* (333)
Puttock	Common Buzzard *Buteo buteo* (70)
Pyrarg	White-tailed Eagle *Haliaeetus albicilla* (61)
Rock Ouzel	Rock Thrush *Monticola saxatilis* (195)
Skout	Common Guillemot *Uria aalge* (324)
Small water-hen	Baillon's Crake *Porzana pusilla* (315)
Solitary sparrow	Blue Rock Thrush *Monticola solitarius* (191)
Water Ouzel	European Dipper *Cinclus cinclus* (149)
Witwall	Eurasian Golden Oriole *Oriolus oriolus* (198)
Woodspite	Green Woodpecker *Picus viridis* (135)

Notes

NB: Mi LM refers to the Middleton Collection, which contains the Francis Willughby archive, held in Manuscripts and Special Collections, the University Library at the University of Nottingham.

CHAPTER 1: BITTEN BY THE SNAKE OF LEARNING

1 These details, including those of the birth of his two sisters, Lettice (17 March 1627) and Katherine (4 November 1630, also referred to as Catherine), were on a page found in a copy of Joseph Hall's *Complete Works* (London, 1628). D. Johnston, pers. comm.
2 https://en.wikipedia.org/wiki/Wollaton_Hall.
3 Cram et al. (2003).
4 Cassandra Ridgeway's date of birth is unknown, but was probably around 1600. She was 'married' (betrothed) to Sir Francis at the age of eleven; he then went travelling and the couple probably did not live together until they moved into Middleton in 1615. (D. Johnston, pers. comm. (2016)). Such early betrothals were not unusual, but marriage at fifteen was (A. Brabcová, 'Marriage in Seventeenth-Century England: The Woman's Story', in *Theory and Practice in English Studies*, eds P. Drábek and J. Chovanec, vol. 2, 21–4, Proceedings from the Seventh Conference of English, American and Canadian Studies, Brno, 2004): https://www.phil.muni.cz/angl/thepes/thepes_02_02.pdf.
5 Boran (2004).
6 Wood (1958).
7 Seth Ward (1654) cited by Serjeantson (2016).
8 Serjeantson (2016).
9 Ibid.

10 Thomas (2002).

11 Ibid.; Serjeantson (2016).

12 Feingold (1990: 31).

13 Ibid. (33), from Charleton's book *The Immortality of the Human Soul, Demonstrated by the Light of Nature in Two Dialogues*. London: William Wilson, 1657.

14 http://www.biography.com/people/francis-bacon-9194632#philosopher-of-science.

15 Wootton (2015: 73).

16 Ibid. (136).

17 'Memoir of Dr James Duport', eds James Henry Monk and Charles James Blomfield. Cambridge: John Murray, 1846, 672–97.

18 Feingold (1990).

19 Ibid.

20 Ibid.

21 Oswald and Preston (2011).

22 Wood (1958).

23 Yeo (2014).

24 Serjeantson (2016).

25 Oswald and Preston (2011).

26 Muffett – see Chapter 7.

27 From Duport on Willughby, *Musae subseciave*, translated by Philip Oswald.

28 Serjeantson (2016).

29 Ibid.

30 Broad (1944).

31 Williams (1966).

32 Parker (2006); Birkhead and Monaghan (2010).

33 From a Latin poem by James Duport addressed to Willughby (Duport 1676: 316), translated by Philip Oswald, pers. comm.

34 Feingold (1990: 12).

35 Serjeantson (2016).

36 From a Latin poem by James Duport addressed to Willughby (Duport 1676: 316), translated by Philip Oswald (Birkhead 2016a).

37 Raven (1942: 51–2).

38 Serjeantson (2016).

CHAPTER 2: JOHN RAY AND THE CUNNING CRAFTSMANSHIP
OF NATURE

1 Raven (1942: 62) in his authoritative biography of John Ray states that
   there are just three portraits 'from life' of John Ray: 1. A young, forty-
   ish Ray without a wig, an oil painting thought to have been executed
   around 1667 and attributed to Mary Beale (Natural History Museum,
   London). 2. Ray in his sixties – the best-known oil portrait and by
   an unknown artist (National Portrait Gallery, London). 3. Ray in his
   sixties, chalk and graphite from around 1691 by William Faithorne
   (Department of Prints, British Museum). Jardine (2003), in her biography of Robert Hooke, dismissed the idea that portrait 1 was of Ray,
   suggesting instead that this is Robert Hooke. Jensen (2004) in turn
   dismissed this, proposing instead that the portrait is of Jan Baptist Van
   Helmont (1579–1644). The key piece of evidence in support of this is
   a line drawing of Van Helmont in his own book published in 1644.
   The similarity to the oil painting (portrait 1) is so strong it appears that
   Mary Beale constructed her painting from the drawing. The weight of
   evidence, then, is that portrait 1 is not Ray. Nonetheless, it is intriguing
   that Raven, who was obsessed by Ray, should accept that portrait 1 was
   of his subject, as did Sawyer (1963).
2 Feingold (1990: 2).
3 Raven (1942: 27).
4 Wardhaugh (2016).
5 Mynott (2018).
6 Calloway (2014: 98); Raven (1942: 37).
7 Ogilvie (2016a).
8 B. W. Ogilvie, pers. comm. (10 November 2015).
9 Ray (1691).
10 Ibid., xiii.
11 I erroneously ascribed this idea to John Ray (Birkhead 2008), but
    later found that the idea had come from Robert Boyle – see Ray,
    The Wisdom of God (1735: 122; 10th edition). Lack (1966) studied the
    timing of birds' breeding seasons and showed it to be true.
12 Wootton (2015).
13 Raven (1942: 81). McMahon (2000) suggests that Ray's mental ill-
    ness was caused by distress at the Civil War.
14 Raven (1942: 82).
15 C. D. Preston, pers. comm.

16 Oswald and Preston (2011).

17 Ibid.

18 Raven (1942: 47–8); Feingold (1990: 34).

19 Raven (1942: 313).

20 *Ornithology*, 282.

21 Ibid.

22 Baldner (1653); see facsimile, Baldner (1973); MacGillivray (1852).

23 Feingold (1990: 35).

24 Roos (2016).

25 Lemery (1675).

26 Bynum (2008).

27 Derham (1718: 357–8).

28 https://en.wikipedia.org/wiki/Chemical_garden.

29 So wrote William Holder in 1678. In that same year a certain John Wallis claimed 1645 to be the year of the Society's foundation, but as John Gribbin has pointed out in *The Fellowship* (2005), the dispute about the date arose because Wallis dishonestly took credit for some of Holder's scientific findings and in the ensuing battle contradicted all that Holder had said, including the date on which the Royal Society began.

30 Gribbin (2006: 125–9).

31 Jardine (1843).

32 Johnston (2016).

33 *Ornithology*, Preface, 1.

34 Ibid.

35 Ibid., 2.

36 Wood (1958: 107–11).

CHAPTER 3: A MOMENTOUS DECISION

1 Burnet (1833).

2 Gribbin (2006: 133).

3 Hooke (1665), Preface to *Micrographia*.

4 Ogilvie (2016b).

5 The Royal Society's repository: see Hunter (1995).

6 *Ornithology*, 334, and plate 78.

7 Both Keynes (1964) and Barbour (2013: 436) allude to Browne's daughter making illustrations for him, but I could find nothing definitive.

8 Raven (1942: 116).

9 Ibid. (115).

10 Ray (1942); Lankester (1846: 165).

11 Ibid. (Lankester: 250 n. 3).

12 Plot (1686: 229).

13 Gurney (1921: 189); Shrubb (2013).

14 Ray (1942); Lankester (1846: 166).

15 Ray (1942); Lankester (1846: 168).

16 Cram and Awbery (2001). We know that Willughby kept notebooks (now lost) for journeys other than Prestholme because he refers to them elsewhere. These include a 'Cornwall Journy' mentioned in his Commonplace Book (Mi LM 15/1 p. 481) and a 'Worcester Journy' mentioned in a letter (Mi LM 15/1 p. 486); doubtless there were others (D. Johnston, pers. comm.).

17 Derham (1760).

18 See Appendix 1.

19 Ray (1686a).

20 Lankester (1846: 171).

21 Ibid. (173).

22 Buxton and Lockley (1950); Birkhead (2016b); and see http://www.welshwildlife.org/wp-content/uploads/2011/05/History-of-Skomer-including-timeline.pdf.

23 *Ornithology*, 18.

24 Birkhead (2016c).

25 *Ornithology*, 18; Harvey (1981: 66).

26 Lankester (1846: 154).

27 http://en.wikisource.org/wiki/Willisel,_Thomas_(DNB00), and Raven (1942: 151). Thomas Willisel was employed by the Royal Society to search out natural rarities: Ray described him 'as the fittest man for such a purpose that I know in England' (Lankester 1846: 151–2).

28 Ray and Willughby examined specimens of and described the red grouse in 1668 – see Chapter 7.

29 *Ornithology*, 326–7.

30 Lankester (1846: 175).

31 Ibid. (176).

32 This is based on the account given by Ray to his friend William Derham towards the end of Ray's life, when on 15 May 1704 Derham 'waited upon him at Black Notley' and wrote: 'These two gentlemen [Willughby and Ray], finding the History of Nature very imperfect, had agreed between themselves, before their travels

beyond the sea, to reduce the several tribes of things to a method; and to give accurate descriptions of the several species, from a strict view of them. And forasmuch as Mr Willughby's genius lay chiefly to animals, therefore he undertook the birds, beasts, fishes, and insects, as Mr Ray did the vegetables [plants]' (Lankester 1846: 33). Here, 'insects' comprise true insects, but also a number of other invertebrate groups. What seems to be missing from Willughby and Ray's subsequent writings are the 'beasts' or what Willughby and his contemporaries also called quadrupeds. It is unclear why quadrupeds (or rather, mammals) – with the exception of cetaceans – get almost no notice in Willughby and Ray's work. Only in the 1690s and long after Willughby's death did Ray acknowledge cetaceans to be distinct and more similar to quadrupeds than fish (Romero 2012). In the Preface (p. 3) to the *Ornithology*, Ray refers to Willughby's interest in 'beasts': 'Viewing his manuscripts after his death, I found the several animals in every kind both birds, beasts, fishes and insects digested into a method of his own contriving', but apart from a few images of mammals in the Middleton Collection, there is little evidence that Willughby planned a volume on mammals.

33 Lankester (1846: 177).

34 Cram and Awbery (2001); Cram (2016).

35 From Willughby's commonplace book: D. Johnston, pers. comm.

36 Diary of John Evelyn, p. 292, of this online version: https://archive. org/stream/diaryofjohnnevely01eveliala.

37 Derham (1718).

38 The name 'snap-apple bird' used here by Willughby may have come from Richard Carew's *Survey of Cornwall* (1602) – which Willughby owned – in which he notes the occurrence of a 'flocke of birds in biggnesse not much exceeding a sparrow which made a foule spoyle of the apples. Their bils were thwarted crosse-wise at the end, and with these they would cut an apple in two at one snap eating only the kernels [seeds]' (cited in Raven 1947: 246). In the *Ornithology* (248), Willughby and Ray refer to the crossbill as the 'shell-apple', a name Christopher Merret used in his ornithological compilation, *Pinax* (1667), as a name for the crossbill. This is a little confusing since Turner (1544) had previously used 'sheld-appel' to refer to the chaffinch.

39 Hunter (1995: 37).

40 R. Serjeantson, pers. comm. The drainage of the Lincolnshire Fens by Dutch engineers, instigated by Charles I, was already underway by 1660, and was causing the loss of livelihoods for fishermen and wildfowlers.

41 Lankester (1846: 185–8).

42 McMahon (2000).

43 R. Serjeantson, pers. comm; McMahon (2000).

44 McMahon (2000).

45 Raven (1942: 61).

46 Johnston (2016: 7). The same letter as that in n. 37, but in its complete form, in the Middleton Collection.

47 From Birch (1756–7, vol. I, 114): The observation of this embryo snake's external hemipenes *might* have constituted a discovery had Willughby provided more eggs and from more clutches showing the same phenomenon among the male embryos, *and* noted that the hemipenes of newly hatched male snakes of the same species were no longer external, but were held inside the cloaca. It appears that no one (including Aristotle, Aldrovandi, Gessner and Topsell) had noted the external hemipenes of embryo male snakes (M. Olsson, pers. comm., K. Adler, pers. comm., November/December 2016).

48 Willughby attended Royal Society meetings on 1, 8 and 15 October 1662. On 8 October he presented his quincuncial tree-planting solution; on 15 October he demonstrated the skin covering the eye of a whiting (fish), and continued a discussion of snakes.

49 From Hunter (1995).

CHAPTER 4: CONTINENTAL JOURNEY: THE LOW COUNTRIES

1 Skippon (1732: 361).

2 Greengrass (2016).

3 Ibid.; Howell (1642).

4 Full quote: 'those base and badder minds ... who content their poore thoughts with their owne countries knowledge ... were like sillie birds cooped up in a pen'. We have no direct evidence but some of Lipsius's other works were in the Willughby library: W. Poole, pers. comm., and Poole (2016).

5 Iliffe (1998); Zeiler (1656).

6 For details of naval tensions see Greengrass (2016).

7 Iliffe (1998). Ray does not comment on Kircher's museum in his published account of their travels, but the recent discovery of some of Ray's manuscripts shows that he and Skippon visited the museum while travelling separately from Willughby and Bacon. The reason Ray omitted any reference to the museum may have been because he disapproved of the 'self-promotion that was so central a feature of Kircher's attitude to his museum' (Hunter 2014).

8 Greengrass (2016).

9 Kusukawa (2016); see also Kusukawa (2000).

10 Hunter (2014).

11 Zwinger (1577).

12 Ray (1738).

13 Ray (1713: 150).

14 Ray (1738: 18). In Antwerp, Ray listed the many rare plants in the garden of the priest Franciscus van Steerbeck.

15 Ibid. (71).

16 Skippon (1732: 380).

17 Storey (1967).

18 Ray (1738: 19).

19 Skippon (1732: 385).

20 *Ornithology*, 61.

21 Ray (1738: 19) and Skippon (1732: 385).

22 *Ornithology*, 88.

23 Ibid., 87–8.

24 Ibid., 89.

25 Skippon (1732: 389).

26 Ray (1738: 24).

27 Ibid.

28 Ibid. (27–30).

29 Ibid. (31).

30 Ibid. (32) and Skippon (1732: 400).

31 Skippon (1732: 399–400).

32 Ray (1732: 27). Descartes on pineal gland: Lokhorst (2016).

33 *Ornithology*, 332. In Britain in Willughby's day cormorants almost invariably nested on cliffs or on the ground rather than in trees, but the continental race of the cormorant *Phalocrocorax carbo sinesis* – the one Willughby and Ray saw in the Netherlands – did/does nest

in trees: hence their surprise. In recent years *P. c. sinesis* has been recorded in Britain, and now some of 'our' race of the cormorant also nest in trees (Newson et al. 2007). On sending two boatloads to England, see Hegenitius and Ortelius (1630).

34 Graham Martin, pers. comm.

35 *Ornithology*, 288–9.

36 Preston (2016), an appendix to Greengrass (2016): the letter from Willughby to Ray was written in early September 1661, and I have assumed that Willughby was then at Middleton.

37 Mabey (2015).

38 The idea that skinks could be taken as an aphrodisiac originated with Pliny (*Natural History* XXVIII, 119), who states that 'The Indian is the biggest *scincus,* then the Arabian one. They import these salted. Its snout and feet, taken with white wine, are aphrodisiac, especially when mixed with satyrion [either an orchid or ragwort] and rocket seed, mixing one handful of these with two handfuls of pepper.' This sounds like a powerful concoction, but I have been unable to find any evidence for its efficacy.

39 A feral colony of night herons exists or existed near Edinburgh – escapees from Edinburgh Zoo (Dorward 1955) – but at the time of writing (2017) this 'population' comprises a single geriatric individual. A wild pair bred in Somerset in 2017.

40 *Ornithology*, 279.

41 Ibid., 277. I found it hard to believe that no one had previously seen and commented on Willughby's collection of eggs, and sure enough, just when I had finished writing, I was reading John Gurney's *Early Annals of Ornithology* (1921), a much undervalued book on the history of ornithology, and found (p. 211) the following: 'The remnants of his egg collection still exist in a cabinet at Wollaton Hall. Most of the eggs have been written upon, and the writing is still legible, although the eggs themselves are faded and mostly cracked. Some heron's eggs remain intact and these and an inscribed shoveler duck's egg may have been procured by Willughby or Ray in Warwickshire, but we are tempted to associate a night heron's egg (?) marked "Quacke Belge" with their visit to Sevenhuis in Holland.' Interestingly, although the existence of the zoological specimens is poorly known, the seed collection has been known for some time (Welch 1972).

42 Vestjens (1975).

43 Raven (1942: 317).

44 *Ornithology*, 136.

45 Roos (2015: 446).

46 Ray (1738: 48).

47 http://www.historyofparliamentonline.org/volume/1604-1629/member/barnham-sir-francis-1576-1646.

48 Raven (1942: 65); Roos (2015: 118). A contemporaneous recipe for a tansy (an egg pudding) involved fifteen egg yolks, some sugar, dry sack (Spanish wine), a pint of cream, a flavouring of tansy, spinach and primrose leaves, all cooked until 'pretty stiff' and then fried in sweet butter (Samuel Pepys, *The Pepys Companion*).

49 Ray (1738: 44).

CHAPTER 5: IMAGES OF CENTRAL EUROPE

1 We know nothing of the meeting between Willughby and Balder: I have assumed, on the basis of information in Phillips (1925: 333) and Fluck and Scharback (2016), what took place.

2 Raven (1942: 354).

3 Phillips (1925: 336).

4 The quote is from Phillips (1925); Birkhead (2014: 131); Whitten (1971).

5 Phillips (1925).

6 Annotations to the drawing of a cormorant in the Brown University copy of Baldner.

7 Johannes Faber (1649), cited in Beike (2012).

8 *Ornithology*, 332.

9 Raven (1942: 65); Roos (2015: 118). A contemporary recipe for a tansy (an egg pudding) involved fifteen egg yolks, some sugar, dry sack (Spanish wine) a pint of cream and a flavouring of tansy, spinach and primrose leaves, all cooked until 'pretty stiff' and then fried in sweet butter (Latham and Matthews 1983: 148).

10 Phillips (1925: 338).

11 *Ornithology*, Preface, 6. Ray's reference to Baldner as 'this poor man', reinforces my view that Ray did not meet Baldner, and that Willughby went alone to meet him (see n.1 above), since the study by Fluck and Scharbach (2016) shows that Baldner was 'not an ordinary fisherman with a gift for nature study, but a highly valued member of the fishermen's corporation, a wealthy, learned citizen and a considerate father. He bequeathed a considerable legacy

to his wife and children.' They refer to him as a 'fisherman and magistrate'.

12 Lauterborn (1903).

13 Phillips (1925).

14 Notes quoted from Brown University web pages; 5,000 books bequeathed to Brown in 1979: https://search.library.brown.edu/ ?utf8=%E2%9C%93&q=Baldner.

15 Lownes (1940).

16 From Lauterborn (1903).

17 Anon.

18 Skippon (1732: 459).

19 *Ornithology*, 131.

20 Koreny (1985).

21 See Ray (1738). Philip Skippon's description of Florence in https:// archiv.ub.uni-heidelberg.de/artdok/1216/2/Daly_Davis_Fontes51.pdf.

22 *Ornithology*, 147.

23 Clark (1981).

24 Skippon (1732: 468); Muffet in *Health's Improvement* (1655) does not mention raptors being eaten at all, but of jays he says that they 'never come unto the number of good nourishment'; of the starling he says that the 'flesh is dry and sanery [meaning not clear]'; of the wren, 'no man ever wrote that it give good nourishment', and of tits, 'all of them feed ill, and nourish worse'. Skippon may have obtained his opinion about which birds were worth eating or not from Muffet.

25 *Ornithology*, Preface, 6.

26 Skippon (1732: 468).

27 Ibid.

28 Hoffman (1660).

29 Skippon (1732: 469).

30 Throughout the Middle Ages and right up to Willughby's day this species was known as St Cuthbert's duck. 'Imprinted' refers to the fact that it was hatched in captivity and as a result 'imprinted' on its owner, St Cuthbert, because this was the first thing it saw.

31 Skippon (1732: 470).

32 Greengrass (2016).

33 Aldrovandi (1648).

34 Roos (2016).

35 Mynott (2018).

36 Eyles (1955); Edwards (1967).

37 Skippon (1732: 476–9).
38 On 1 August 1664 – known as the Battle of St Gotthard in Ottoman and Hungarian sources, and as the Battle of Mogersdorf in Austrian ones.
39 Skippon (1732: 480).
40 Ibid.
41 Ibid. (481).

CHAPTER 6: ITALIAN SOPHISTICATION AND SPANISH DESOLATION

1 Ray (1738: 124).
2 Zimmerman (2008): water from snow or glaciers lacks iodine.
3 Ray (1732: 168).
4 Skippon (1732: 502): literally *Out, out* – onto the stage; in modern Italian: *Fuori*.
5 Ibid.
6 Greengrass (2016).
7 *History of Insects*, 7, trans. John Wade (pers. comm.).
8 Skippon (1732: 500).
9 Ibid. (501); Broise and Jolivet (2009). Note: there are plenty of current North American recipes for terrapin soup or stew on the Internet, even though most freshwater turtles are protected.
10 Skippon (1732: 519).
11 Ibid. (496).
12 *Historia Piscium*, 132.
13 Ibid. (307).
14 Ibid. (287).
15 Ibid. (156).
16 Ibid. (294).
17 Ibid. (300).
18 Ibid. (311).
19 Ibid. (331).
20 Ibid. (132).
21 Skippon (1732: 504).
22 MacPherson (1897: 35); Vogelsang (1949).
23 Angelini (1724).
24 *Ornithology*, Preface, 6.
25 Ibid. (196).
26 Jackson (1993, vol 1: 42).

27 I asked the Caravaggio expert Andrew Graham Dixon about this and he was confident that the painting was not by Caravaggio (pers. comm.). On the website below it is attributed to the Master of Hartford; it is in the Galleria Borghese, Rome, dated before 1607. http://www.gettyimages.co.uk/detail/illustration/still-life-with-birds-by-master-of-hartford-1590-1600-stock-graphic/132701607.

28 Skippon (1732: 517).

29 T. Pizzari, pers. comm.

30 Skippon (1732: 534). Note: £25 in the 1660s was very roughly equivalent to £1,700 today; and the ten zucchini offered by Willughby was approximately half of what Regio asked for. Tom Pizzari found this text, *Nuova geografia, tradotta in lingua toscana da Gaudioso Jagemann* (1780), from a library in Vienna. Volume 15 has a footnote which implies that 1 zecchino was worth £0.5. They met Regio in Padua.

31 Gessner, cited in Smith and Findlen (2001).

32 Ibid.

33 Park (1997).

34 Ray (1738: 182).

35 Hunter (2014).

36 Preston (2016).

37 Technically an amphipod, not a true shrimp, and related to the more familiar *Gammarus* of British streams and rivers, *Niphargus costozzae* occurs in many cave systems in the Vento and Verona regions of northern Italy. It was formally identified and named in 1835 (Pierre Marmonier, pers. comm.). See https://ortobotanicobologna.wordpress.com/ovidio-montalbano-en.

38 Ray (1738: 188).

39 Skippon (1732: 573).

40 See https://ortobotanicobologna.wordpress.com/ovidio-montalbano-en.

41 See http://www.filosofia.unibo.it/aldrovandi/pinakesweb/main.asp?language=it.

42 Skippon (1732: 757).

43 Greengrass (2016).

44 https://valentinagurarie.wordpress.com/tag/conrad-gessner.

45 Ragnar Kinzelbach, pers. comm.

46 Ray (1738: 234). The Academy visited by Willughby in Naples was L'Accademia degli Investiganti (Researchers or Investigators Academy), founded in 1630 and shut down in 1656 because of the plague; reopened in 1664. The Academy was shut down again by the viceroy in 1668, although members continue to meet informally after this. Several prominent members were arrested because of their openly secular views, but narrowly escaped the Inquisition thanks to an earthquake: https://it.wikipedia.org/wiki/Accademia_degli_ Investiganti.

47 Ray (1738: 52, 136).

48 Raven (1942: 138); Lister had been at St John's in Cambridge while Ray was at Trinity, so it is possible that they already knew each other before meeting in the sunny south of France.

49 Lewis (2012).

50 Birkhead and Charmantier (2009).

51 Ray (1738).

52 Howell (1642: 52); Zeiler (1656).

53 Derham (1718).

54 *Ornithology*, 315.

55 Derham (1718: 12): the original is in the Natural History Museum archives, MSS Ray, fo. 2, item 3.

CHAPTER 7: BACK AT MIDDLETON

1 Wood (1958: 92).

2 He died on 7 December 1665 and like other family members was duly buried at Middleton church.

3 Wood (1958: 110).

4 Ibid. (108).

5 Ibid. (109).

6 Letter dated 22 October 1666 in Derham (1718).

7 Oswald and Preston (2011: 276).

8 Ibid. (278).

9 Turner (1554) cited in Evans (1903); Oswald and Preston (2011: 276).

10 Aristotle, *History of Animals*, Book V, Part 1.

11 Oswald and Preston (2011: 276).

12 Willughby (1671).

13 Ibid.; Lister (1671).

14 Lister (1671).

15 The parasitoids that Willughby observed in the cabbage white larvae were the larvae of the wasp *Cotesia glomeratus*: see Thomas and
Lewington (2010).

16 Ray, *Historia Insectorum* (1710), cited in Oswald and Preston
(2011: 279).

17 Ibid. (279); see also Lentern and Godfray 2005 (cited therein).
Leeuwenhoek's first letter on aphid parasitoids to the Royal
Society is dated 1687 (Malina 2008). We also know that
Willughby owned a copy of Goedaert's *Metamorphosis Naturalis*
(1660–9), which contains images of parasitoids, as well as differ-
ent life stages of insects.

18 Letter from Charles Darwin to Asa Gray 22 May 1860. Darwin
Correspondence Project, 'Letter no. 2814', accessed 2 January 2017.

19 King (1614–81) was identified by Roos (2015: 445 n. 3).

20 Birch (1756–7), vol. II, pp. 435 and 449, on Willughby's two letters
being read to the Royal Society.

21 King (1670).

22 Quoted in Jardine (1864: 87).

23 Willughby (1671) cited in Ogilvie (2016a).

24 Mullens (1911) says Spain rather than Italy.

25 From Neri (2011: 31, 202).

26 Muffet (1634).

27 It is not known whether Thomas Muffett wrote the rhyme for
her; it first appears in print in the early 1800s (Opie and Opie
1997: 323–4).

28 Raven (1942: 183).

29 Roos (2015: 230).

30 As Romero (2012) has pointed out, when Ray was in Chester in
1669 dissecting his porpoise, he considered it to be a 'fish' (albeit a
special one); by the time he produced *Historia Piscium* in 1686, he
realised that cetaceans are distinct, and it is only in 1693, when he
produced *Synopsis Methodica Animalium Quadrupedum et Serpenti
Generis*, that he was completely convinced that cetaceans were not
fish. He says: 'For except as to the place on which they live, the
external form of the body, the hairless skin and progressive swimming motion: they have almost nothing in common with fishes, but
remaining characters agree with the viviparous quadrupeds.'

31 Roos (2015: 457). The death of John Wilkins: he was thought initially to have died 'of the stone', that is kidney stones, hence the

'stoppage of urine', but an autopsy revealed only 'two small stones in one kidney and some litte gravell in one ureter but neither big enough to stop the water. Twas believed his opiates and some other medicines killd him' (from the diary of Wilkins's close friend Robert Hooke, cited by Jardine [2003]).

32 Locke (1690), D. Cram, pers.comm.

33 Newton (1896: 31).

34 Stresemann (1975: 177).

35 Firestein (2012).

36 The differences between the honey-buzzard and common buzzard are from the *Ornithology*, p. 72; the rest is my reconstruction.

37 Roos (2015: 98).

38 Raven (1942: 143–5).

39 Wood (1958); Johnston (2016). We do not know for certain who Thomas Alured was, but the most likely candidate is the son of John Alured (1607–51), Parliamentary officer and regicide. Thomas Alured entered Gray's Inn on 12 July 1655 and was therefore roughly contemporaneous with Francis Willughby, who was admitted in 1657, and with Francis Jessop, admitted in 1656. The Alured family lived in or near Hull and Thomas Alured became a barrister and a recorder (1688–1700) of Beverly (near Hull) and gave a large collection of books to Beverly parish library; http://www.robert-temple. com. Assuming Thomas Alured was a friend of Willughby's, this may explain Francis's ownership and annotation of one of the printed records of the legal action against the regicides: *An exact and most impartial accompt of the indictment, arraignment, trial, and judgment (according to law) of nine and twenty regicides, the murtherers of His late sacred Majesty of most glorious memory* (London: Printed for Andrew Crook and Edward Powel, 1660).

40 Johnston (2003). Katherine married (in 1666) Clement Winstanley, a 'gentleman that loved his pleasures, and spent more to gratifie them than his estate would bear'. He died, in debt, in 1672 (Wood 1958).

41 Wood (1958: 111).

42 Raven (1942: 316, 319); Armytage (1952).

43 Ragnar Kinzelbach, pers. comm.

44 *Ornithology* (259); Derham (1718: 367).

45 Derham (1718: 367).

46 Roos (2015: 176).

47 Derham (1718: 368).

48 I suspect that the impetus for Francis's experiments on the flow of sap in the spring of 1665 was the Royal Society's committee, established earlier that year, whose 'Queries concerning vegetation' were eventually published in 1668. The information was transcribed from Francis Willughby's comonplace book (Mi LM 15 pp. 433–4) by Dorothy Johnston and sent to the author on 15 November 2016.

49 The Royal Society's paper was: 'Queries concerning vegetation, especially the motion of the juyces of vegetables' (1668); Willughby and Ray's brief account was published in the Royal Society's *Philosophical Transactions* in January 1669.

50 Roos (2015).

51 http://5e.plantphys.net/article.php?id=99.

52 Johnston (2003).

53 Hunter (1989: 247–9).

54 Hall and Hall (1965–86): letters starting 19 March 1670.

55 Ibid.

56 H. Smith, pers. comm. However, wolf spiders (*Lycosidae*) have been reported attacking tiny, newly metamorphosed American toads *Anaxyrus* (*Bufo*) *americanus* (*Ecology* 2014: 95: 1724–30). Antoni van Leeuwenhoek (he who built his own microscopes and discovered the existence of spermatozoa) reported to the Royal Society in 1701 that he had conducted experiments in which he set up spiders with young frogs (about one and a half inches in length), one of which died after being bitten by 'a great spider'.

57 Welch (1972).

58 Cram et al. (2003: 61).

59 Ibid. (198).

60 Ibid. (71).

61 Hall and Hall (1965–86).

62 https://en.wikipedia.org/wiki/The_Garden_of_Cyrus, and Aldersey-Williams (2015).

63 Cram et al. (2003: 55).

CHAPTER 8: CURIOUS ABOUT BIRDS, ILLNESS AND DEATH

1 Huynen et al. (2010).

2 Jardine (1843).

3 https://en.wikipedia.org/wiki/History_of_the_telescope.

4 Ransome (1947): it seems from Wikipedia that the phrase may have come from an ornithologist and adult fan, Myles North.

5 *Ornithology*, 188.

6 The images in the *Collins Bird Guide* (2009) were originally published by Bonnier Fakta in Sweden; Collins produced it for the English market.

7 Dan Zetterström, pers. comm., 11 June 2015.

8 Fowler et al. (2009).

9 Cade (1967); T. Cade, pers. comm., 9 January 2013.

10 Schulze-Hagen and Birkhead (2015); Birkhead et al. (2014).

11 Leisler and Schulze-Hagen (2013).

12 Those 'functional' questions also re-emerge in Ray's *Wisdom of God* (1691), which was based on his early sermons.

13 Aristotle, *History of Animals*; Aelian: http://penelope.uchicago.edu/aelian/varhist1.xhtml; and Harvey on *Generation* (see Whitteridge, trans., 1981).

14 Durham and Marryat (1908).

15 Thuman et al. (2003); Jaatinen et al. (2010).

16 Ray (1738: 128–9).

17 *Ornithology*, 169 (quail) and 249 (house sparrow – famous for its frequent copulations).

18 Gunn (1912); Fitzpatrick (1934).

19 Kinsky (1971).

20 Feare (1984).

21 Gurney Jr. (1878) refers to guillemot eggs found on the seabed well away from any colony and states: 'We have in our collection an egg that was dredged up at Lowestoft at a depth of 24 fathoms.'

22 Reynard and Savory (1999).

23 Birkhead et al. (2011: 24); Chaplin (1976); Reinertsen (1983).

24 Belon (1555).

25 Belon (1555).

26 Valli da Todi (1601); Olina (1622); Manzini (1575).

27 Birkhead and Charmantier (2013).

28 Other aspects of bird biology established from studies of captive birds include: ways of distinguishing the sexes without the need for dissection; the underlying mechanisms of birds' breeding seasons; instinct and intelligence; and longevity (Birkhead and van Balen 2008).

29 *Ornithology*, 373.

30 Ibid.

31 Willughby's strange (and hardly comprehensible) ideas on moult are cited by Raven (1942: 336) as an example of an outlandish idea that Ray would have rolled his eyes at. It is curious, then, that Ray bothered to include that particular bit of text in the *Ornithology*. Mary Fissel, pers. comm.

32 Stresemann (1975).

33 Brian Follett, pers. comm.

34 Stresemann and Stresemann (1966).

35 Kral (1965).

36 Mounted skeletons (Skippon 1732: 391).

37 Letters dated 28 April and 22 December 1670 (Roos 2015: 266–7 and 285). 'Tertian ague' must have been a general term, for both John Ray and Martin Lister as well as Francis Willughby suffered from occasional bouts of it (Roos 2015: 262–3).

38 Johnston (2016).

39 Ibid.

40 Birch (1756–7).

41 This is from Wood (1958) with information on medical practice in the seventeenth century provided by Michael Collins (British Society for the History of Medicine, pers. comm., 26 October 2016). Willughby owned a copy of Sydenham's (1666) *Methods for Curing Fevers* (William Poole, pers. comm., 26 October 2016). Information on Dr Anthony Hewitt, Willughby's physician, came from Jonathan Barry, pers. comm., 18 October 2016.

CHAPTER 9: INTO THE LIGHT: PUBLICATION

1 Johnston (2016).

2 Katherine Willoughby's husband, who died in 1672, was Clement Winstanley; see Johnston (2003, 2016).

3 Information on the magpie comes from two letters in the Middleton Collection (Mi Av 143/36/20 and 21): D. Johnston, pers. comm.

4 Raven (1942: 320) provides details of these volumes.

5 Poole (2016).

6 Doherty (2010); Birkhead et al. (2016).

7 Raven (1942: 310); Ray said this of Charleton: 'Whatever he may boast of his performance ... he ... did not understand animals nor had any comprehensive knowledge of them.'

8 Stresemann (1975), Charlton at the RS 22 October 1662: 'Dr Charlton [sic] brought in his papers, in which he had reduced birds into certain families, in Latin and English; which papers were ordered to be kept; and the doctor was desired, in conjunction with Dr Merrett, to reduce fishes into the like classes.' (Birch 1756, vol. 1: 118).

9 *Ornithology*, Preface, 7; see also Raven (1942: 310).

10 Volcher Coiter's volume is not on the list in the Willughby library catalogue (W. Poole, pers. comm.); Pierre de La Ramée (aka Petrus Ramus) (1515–72); Theodor Zwinger, author of *Methodus Apodemica*. 'Dichotomous' means dividing or branching into two and is a way of identifying things, such as birds, using contrasting features (e.g. land birds *versus* water birds; webbed feet *versus* not webbed feet; bill yellow *versus* bill not yellow, etc.), into smaller and smaller groups, such that eventually you are left with only two options and therefore are able to identify the species.

11 *Ornithology*, 154.

12 Griffing (2011).

13 *Ornithology*, Preface, 4.

14 Ibid.

15 Griffing (2011).

16 Johnston (2016).

17 Derham abstracted some of Ray's letters containing details of his reproductive issues. His abstracts are in the back of NHM MSS Ray 1 at the Natural History Museum (London), and there is the letter among Lister's correspondence in the Bodleian Library (numbered 0255 in the edition) that gives some of the details. Gunther's *Further Correspondence of John Ray* (1928) includes some of these abstracts. An image of the letter 0255 can be seen at http://tinyurl.com/6wro6r8.

18 Marquis de Sade: https://en.wikipedia.org/wiki/Cantharidin.

19 Heneberg (2016); another idea is that bustards consume blister beetles as a form of self-medication against parasites (Bravo et al. 2014).

20 Raven (1942: 309).

21 Roos (2015: 8).

22 The term 'history', as in 'natural history', meant, at least originally, the results of their own investigations. In contrast, the term *traité* (as in *Traité du rossignol* [nightingale] by Nicolas Venette [1697] – see Chapter 8) referred to an account based on existing literature.

23 Griffing (2011).

24 Roos (2015: 155). Lister's advice was either verbal, or the letter no longer exists (A. M. Roos, pers. comm.).

25 Ray to Henry Oldenburg, 19 September 1674, in Ray (1928), 66. See also Wragge-Morley (2010) on the importance of written descriptions.

26 Montgomerie and Birkhead (2009).

27 Olina (1622).

28 Marcgraf and Piso (1648).

29 The *Pitangaguacul* is the broad-billed flycatcher *Megarhynchus pitan-gua*; the *Yzquauhtli* remains unidentified (Alexander Lees, pers. comm.).

30 Grigson (2016); Flis (2015); Barlow's images were made popular by his illustrated *Aesop's Fables* published in 1666.

31 *Ornithology*, 389.

32 Ibid., 211.

33 Ibid., 209.

34 Ibid., 229.

35 Browne (1646); *Ornithology*, 146.

36 Birkhead et al. (2014).

37 Suh (2016).

38 *Ornithology*, Preface, 6.

39 On the final page of the Preface to the *Ornithology*, Ray lists several errata, and explains that some birds are illustrated twice, because they weren't very well executed the first time.

40 Birkhead (2003); Derham (1718: 140).

41 Roos (2015: 221).

42 *Philosophical Transactions*, 16 August 1669: see Derham (1718: 52–4); see also Roos (2011: 108).

43 Roos (2015: 276 n. 17).

44 *Ornithology*, Preface, 9; 'fatty juyce'; also *Ornithology*, 273.

45 Muffet (1655) mentions eating woodcock brains.

46 *Ornithology*, 290.

47 Muffet (1655). Unfortunately, Hugh Cott's (1945) palatability tests of a wide range of bird species, conducted in the 1940s, are not of much help because his methodology was so poor.

48 Johnston (2016); John Evelyn's *Diary*.

49 Raven (1942: 339); Roos (2015: 745), letter dated 17 July 1676. Ray's botanical output was prodigious – as outlined at length by Raven (1942), securing him the undeniable reputation as the most influential early modern botanist.

50 Raven (1942: 340).

51 Ray (1686b).

52 Ibid, Introduction, 1–25; whales and dolphins, 26–42; sharks and rays, 44–91; bony fishes, 94–343.

53 Raven (1942: 343).

54 Ibid (362); Ray (1686b: 82).

55 Raven (1942: 363–4).

56 Anthony Campbell, pers. comm. (29 January 2016); Boyle (1666).

57 Kusukawa (2000, 2016) and pers. comm. (2 January 2016).

58 Montgomerie and Birkhead (2009). Kate Loveman (pers. comm. 18 October 2017) has suggested to me that this coloured copy of the *Ornithology* may not have been presented to Pepys by Ray since it contains no authorial inscription recording the gift. Rather, she has suggested that it is equally (or more?) likely that Pepys commissioned the colouring himself, or that his partner (from 1670) Mary Skinner, who was known to colour books, either coloured the images herself or commissioned it.

59 Raven (1942: 365); the Monmouth Rebellion, an attempt to overthrow James II.

60 After whom the comet was named.

61 Ogilivie (2016).

62 See http://snailstales.blogspot.co.uk/2010/01/when-snails-were-insects.html.

63 Charmantier et al. (2016: 365).

64 Johnston (2016).

65 Thomas Willoughby was raised to the peerage as Baron Middleton on 1 January 1672, along with others, 'in order to safeguard the ministerial majority of the Lords' (see http://www.historyofparliamentonline.org/volume/1690–1715/member/willoughby-sir-thomas-1672–1729) (D. Johnston, pers. comm.).

66 Charmantier et al. (2016).

67 Ibid.

68 Roos (2017), 'Only meer love of learning': a rediscovered travel diary of the naturalist and collector, James Petiver (c. 1665–1718). *Journal of the History of Collections*. Petiver was elected a Fellow of the Royal Society in 1695. His butterfly book was *Papilionum Britanniae Icones* (1718).

69 Charmantier et al. (2016).

70 Ogilvie (2016b).

71 Mandelbrot (2004); Raven (1942: 305).

72 Ogilvie (2016b).

73 Ibid.

74 Chinery (2012).

75 Wilkins (1668).

76 Birkhead (2008: 40–1); Ray (1691) later suggested there might be an additional 170, making a total of 670 species. There are now thought to be around 10,000 species of birds, although as molecular studies become increasingly refined, there may be many more (Barrowclough et al. 2016).

77 Ray (1691: 119–20).

78 Petiver (1718).

79 Ramsay and Houston (2003).

80 Olgilvie (2016).

81 Kirkby (1802, vol. 1: 10): Willughby's bee is *Megachile willughbiella*.

82 Raven (1942: 366).

CHAPTER 10: A SUSTAINED FINALE: WILLUGHBY'S BUZZARD TAKES FLIGHT

1 Smith (1788).

2 Ibid.

3 Raven (1942: 87).

4 Swainson (1834: 27).

5 Ibid. (30).

6 Ibid. (31).

7 C. T. Wood (1835) and N. Wood (1836); see Birkhead and Montgomerie (2016).

8 N. Wood (1836).

9 The Wood boys were well tutored by Edwin Lankester (see Birkhead and Montgomerie 2016), who also happened later to edit some of John Ray's letters (Lankester 1848).

10 C. T. Wood (1835: 83).

11 N. Wood (1836: 412).

12 Jardine (1843: 43).

13 Newton (1896).

14 Raven (1942: Preface, x).

15 Ibid. (335–6).

16 Ibid. (335). Virtuoso: A virtuoso was an expert academic without an academic position. On this basis Darwin too was a virtuoso. Since then the term has lost much of its lustre, and now often means a mere 'dabbler'. Even in Willughby's day not everyone valued virtuosity – experts were not universally esteemed. Some of the Royal Society's various experiments and demonstrations drew public derision – not least, the blood transfusion (the first in England) performed in November 1667. The procedure was performed on Arthur Coga, a Divinity student suffering from a mild form of insanity, in the hope that introducing some nine ounces of lamb's blood would quell the young man's tempestuous nature. It didn't of course. Thomas Shadwell's play *The Virtuoso*, first performed at the Dorset Garden Theatre in London in 1676, four years after Willughby's death, has been interpreted as a public attack on the Royal Society and some of its less useful experiments – like this one – but in fact the aim of the play seems to have to been satirise those 'virtuosi' who were mere dabblers with no expertise (see Gilde 1970).

17 Raven (1942: 104).

18 Ibid. (336).

19 Ostwald (1909).

20 Branscomb (1985); Woodward and Goodstein (1996).

21 Snow bunting: *Ornithology*, 255; grasshopper warbler: *Ornithology*, 207.

22 Jessop sending Willughby and Ray a scoter: *Ornithology*, 366.

23 *Ornithology*, 365.

24 Ibid. (315). Ray refers to it as 'a small water hen, Aldrovandi calls it *Polipus gallinula minor*'.

25 In a letter to Lister, Ray comments on how difficult raptors are to identify: 'In respect of the genus of falcons and in general of all raptors nature seems to be playing tricks, in that it is well nigh impossible to distinguish precisely between the species. Even the same bird varies the colour of its wings in accordance with its age: in the case

of older birds the colour generally fades and finally degenerates into a whiteish hue.' Roos (2015: 118–19).

26 *Ornithology*, 72.
27 Newton (1896: 426).
28 Birkhead et al. (2018).
29 Ibid.

APPENDIX 3: THE STORY OF WILLUGHBY'S CHARR

1 J. Leland. *The Itinerary of John Leland in or about the Years 1535–43*, ed. L. T. Smith, 1910.
2 J. Ray. *A Catalogue of Freshwater Fish found in England*: an appendix in his *Collection of English Words*. London 1674.
3 T. Pennant. *British Zoology*, vol. III. London 1769.
4 A. Günther. *Contribution to the Knowledge of British Charrs*. Proceedings of the Zoological Society of London 1862: 37–54.

Bibliography

Aldersey-Williams, H. *The Adventures of Sir Thomas Browne in the 21st Century*. London: Granta (2015).

Aldrovandi, U. *Musaeum Metallicum*. Bologna (1648).

Angelini, G. B. *La discrizione dell'uccellare col roccolo*. Bergamo (1724).

Anon. *D'après Nature. Chefs-d'oeuvre de l'Art naturaliste en Alsace de 1450 à 1800*, in Schongauer, Baldung, Grünewald, Weiditz, Walter, eds, catalogue of the exhibition *Chefs-d'oeuvre d'après nature*. Strasbourg: Creamuse (1994).

Armytage, W. H. G. 'Francis Jessop, 1639–1691: A Seventeenth-Century Sheffield Scientist', *Notes and Queries* 197 (1952), 370–1.

Baldner, L. *Vogel, Fisch- und Thierbuch* (1666). Stuttgart: Verlag Müller und Schindler 1973.

Barbour, R. *Sir Thomas Browne: A Life*. Oxford: Oxford University Press (2013).

Barrowclough, G. M., J. Cracraft, J. Klicka and R. M. Zink. 'How Many Kinds of Birds Are There and Why Does it Matter?' *PLoS ONE* 11(11): e0166307. doi:10.1371/journal.pone.0166307. 2016.

Beike, M. 'The History of Cormorant Fishing in Europe', *Vogelwelt* 133 (2012), 1–21.

Belon, P. *L'Histoire de la nature des oyseaux*. Paris: Guillaume Cavellat Gilles Corrozet (1555).

Birch, T. ed. *The History of the Royal Society of London*, 4 vols. London: Millar (1756–7).

Birkhead, T. R. *Bird Sense: What It's Like to Be a Bird*. London: Bloomsbury (2014).

—— 'Changes in Numbers of Common Guillemots on Skomer since the 1930s', *British Birds* 109 (2016b), 651–9.

—— *The Most Perfect Thing: Inside (and Outside) a Bird's Egg.* London: Bloomsbury (2016c).

—— *The Red Canary.* London: Weidenfeld and Nicolson (2003).

—— *The Wisdom of Birds.* London: Bloomsbury (2008).

—— ed. *Virtuoso by Nature: The Scientific Worlds of Francis Willughby FRS (1635–1672).* Leiden: Brill (2016a).

Birkhead, T. R. and B. van Balen. 'Bird-keeping and the Development of Ornithological Science,' *Archives of Natural History* 35 (2008), 281–305.

—— and I. Charmantier. 'Nicolas Venette's *Traité du rossignol* (1697) and the Discovery of Migratory Restlessness', *Archives of Natural History* 40 (2013), 125–38.

—— and P. Monaghan. 'Ingenious Ideas: A History of Behavioral Ecology', in *Evolutionary Behavioral Ecology*, ed. D. F. Westneat and C. Fox. Oxford: Oxford University Press (2010), 3–15.

—— and R. Montgomerie. 'A Vile Passion for Altering Names: The Contributions of Charles Thorold Wood Jun. and Neville Wood to Ornithology in the 1830s', *Archives of Natural History* 43 (2016), 221–36.

—— I. Charmantier, P. J. Smith and R. Montgomerie. 'Willughby's Buzzard: Names and Misnomers of the Honey-Buzzard, *Pernis apivorus.'* *Archives of Natural History* 45 (2018)', 80–91.

—— N. Hemmings, C. N. Spottiswode, O. Mikulica, C. Moskát, M. Bán and K. Schulze-Hagen. 'Internal Incubation and Early Hatching in Brood Parasitic Birds', *Proceedings of the Royal Society of London B* 278 (2011), 1019–24.

—— P. Smith, M. Doherty and I. Charmantier. 'Willughby's Ornithology', in *Virtuoso by Nature: The Scientific Worlds of Francis Willughby FRS (1635–1672)*, ed. T. R. Birkhead. Leiden: Brill (2016), 268–304.

—— J. Wimpenny and R. Montgomerie. *Ten Thousand Birds: Ornithology since Darwin.* Princeton, NJ: Princeton University Press (2014).

Boran, E. 'Hill, William (1618–1667)', *Oxford Dictionary of National Biography*, Oxford: Oxford University Press (2004). http://www.oxforddnb.com/view/article/13315, accessed 4 January 2017.

Boyle, R. 'New Experiments Concerning the Relation between Light and Air (in Shining Wood and Fish); Made by the Honourable Robert Boyle, and by Him Addressed from Oxford to the Publisher, and So Communicated to the Royal Society', *Philosophical Transactions* 2 (1666), 581–600.

Branscomb, L. M. 'Integrity in Science', *American Scientist* 73 (1985), 421–3.

Bravo, C., L. M. Bautista, M. Garcia-Paris, G. Blanco and J. C. Alonso, 'Males of a Strongly Polygynous Species Consume More Poisonous Food than Females'. *PLoS ONE* 9(10): e111057. http://dx.doi.org/10.1371/journal.pone.0111057.

Broad, C. D. 'New Philosophy: Bruno to Descartes', *The Cambridge Historical Journal* 8 (1944), 22–54.

Broise, H. and V. Jolivet. *La Villa Médicis et le Couvent de la Trinité-des-Monts à Rome*. Rome: École Française de Rome (2009).

Browne, T. *Pseudodoxia Epidemica or Enquiries into very many received tenets and commonly presumed truths*. London (1646).

Burnet, G. *Lives, Characters, and an Address to Posterity*, 2nd edn. London: James Duncan (1833), 311.

Buxton, J. and R. M. Lockley. *Island of Skomer*. London: Staples Press (1950).

Bynum, W. T. *The History of Medicine: A Very Short Introduction*. Oxford: Oxford University Press (2008).

Cade, T. 'Ecological and Behavioral Aspects of Predation by the Northern shrike', *Living Bird* 6 (1967), 43–86.

Calloway, K. *Natural Theology in the Scientific Revolution: God's Scientists*. London: Pickering & Chatto (2014).

Carter, J. and J. Basire. *Plans, Elevations, Sections and Specimens of the Architecture of the Cathedral Church of Gloucester*. London: Society of Antiquaries (1807).

Chamberlayne, E. *Angliae Notitia or the Present State of England*. London (1676).

Chaplin, S. B. 'The Physiology of Hypothermia in the Black-Capped Chickadee (*Parus atricapillus*)', *Journal of Comparative Physiology* 112 (1976), 335–44.

Charleton, W. *Onomasticon Zoicon*. London: J. Allestry (1668).

Charmantier, I. and T. R. Birkhead. 'Willughby's Angel: The Pintailed Sandgrouse (*Pterocles alchata*)', *Journal for Ornithology* 149 (2008), 469–72.

Chinery, M. *Insects of Britain and Western Europe*, 3rd ed. London: Bloomsburg (2012).

Clark, G. A. 'Toe Fusion in Oscines', *Wilson Bulletin* 93 (1981), 67–76.

Cott, H. 'The Edibility of Birds', *Nature* 156 (1945), 736–7.

Cram, D. 'Francis Willughby and John Ray on Words and Things', in *Virtuoso by Nature: The Scientific Worlds of Francis Willughby FRS (1635–1672)*, ed. T. R. Birkhead. Leiden: Brill (2016), 244–67.

—— and G. Awbery. 'Francis Willughby's Catalogue of Welsh Words' (1662), *Cylchgrawn Llyfrgell Genedlaethol Cymru/National Library of Wales Journal* 32 (2001), 1–55.

—— J. L. Forgeng and D. Johnston. *Francis Willughby's Book of Games*. Aldershot: Ashgate (2003).

Dale, S. and J. Ray. 'A Letter from Mr Samuel Dale to Dr Hans Sloane, R.S. Secr. giving an Account of what Manuscripts were left by Mr John Ray, together with some Anatomical Observations made at Padua by the said Mr Ray', *Philosophical Transactions of the Royal Society* 25 (1706), 2,282–303.

Derham, W. *Philosophical Letters between the late learned Mr Ray and several of his Ingenious Correspondents, Natives and Foreigners. To which are added those of Francis Willughby Esq*. London: William and John Innys (1718).

—— *Select remains of the learned John Ray, M.A. and F.R.S. With his life*. London: Scott (1760).

Doherty, M. C. *Carving Knowledge: Printed Images, Accuracy, and the Early Royal Society, London*. PhD diss., University of Wisconsin-Madison (2010).

Dorward, D. F. 'The Night-Heron Colony in the Edinburgh Zoo', *Scottish Naturalist* 69 (1955), 32–6.

Duport, J. *Musae subsecivae, seu poetica stromata. Ad. Franciscum Willughbeium, Armigerum, Regiae Societatis Sodalem*, in Liber III (*Sylvarum*), 315–17, in *Obitum Francesci Willughbei, Armigeri, Regalis Societatis Scoii, Pupilli olim sui charissimi*, in *Epicedia*. Cambridge: J. Hayes (1676), 495–6.

Durham, F. and D. C. E. Marryat. 'Note on the Inheritance of Sex in Canaries', *Royal Society of London Report to the Evolution Committee* 4 (1908), 57–60.

Edwards, W. N. *The Early History of Palaeontology*. London: British Museum (Natural History) (1967).

Evans, A. H. *Turner on Birds*. Cambridge: Cambridge University Press (1903).

Eyles, J. M. 'John Ray, F.R.S. (1627–1705)', *Nature* 175 (1955), 103–5.

Feare, C. J. *The Starling*. Oxford: Oxford University Press (1984).

Feingold, M. 'Isaac Barrow: Divine Scholar, Mathematician', in *Before Newton: The Life and Times of Isaac Barrow*, ed. M. Feingold. Cambridge: Cambridge University Press (1990), 1–104.

Findlen, P. 'Natural History', in *The Cambridge History of Science, volume 3: Early Mordern Science*, ed. K. Park and L. Daston. Cambridge: Cambridge University Press (2006), 435–68. http://dx.doi.org/ 10.1017/CHOL9780521572446.020, accessed 2008.

Firestein, S. *Ignorance: How it Drives Science*. Oxford: Oxford University Press (2012).

Fitzpatrick, F. L. 'Unilateral and Bilateral Ovaries in Raptorial Birds', *Wilson Bulletin* 46 (1934), 19–22.

Flis, N. 'Francis Barlow, the King's Birds, and the Ornithology of Francis Willughby and John Ray', *Huntington Library Quarterly* 78 (2015), 263–300.

Fluck, H.-R. and A. Scharbach. 'Leonhard Baldner – Zu seinem Testament und Nachlassverzeichnis', *Revue d'Alsace* 142 (2016), 283–97.

Fowler, D. W., E. A. Freedman and J. B. Scanella. 'Predatory Functional Morphology in Raptors: Interdigital Variation in Talon Size is Related to Prey Restraint and Immobilization Technique', *PLOS ONE* 4(11): e7999. doi:10.1371/journal.pone.0007999 (2009).

Gilde, J. M. 'Shadwell and the Royal Society: Satire in the Virtuoso', *Studies in English Literature 1500–1900* 10 (1970), 469–90.

Goedaerts, M. D. *Metamorphosis Naturalis*, 3 vols. Middelburgh: Jaques Fierens (1662–9).

Greengrass, M., D. Hildyard, C. D. Preston and P. J. Smith. 'Science on the Move: Francis Willughby's Expeditions', in *Virtuoso by Nature: The Scientific Worlds of Francis Willughby FRS (1635–1672)*, ed. T. R. Birkhead. Leiden: Brill (2016), 142–226.

Gribbin, J. *The Fellowship: The Story of a Revolution*. London: Allen Lane (2006).

Griffing, L. R. 'Who Invented the Dichotomous Key? Richard Waller's Watercolors of the Herbs of Great Britain', *American Journal of Botany* 98 (2011), 1,911–23.

Grigson, C. *Menagerie: The History of Exotic Animals in England*. Oxford: Oxford University Press (2016).

Grouw, K. van. *The Unfeathered Bird*. Princeton, NJ: Princeton University Press (2013).

Gunn, T. E. 'On the Presence of Two Ovaries in Certain British Birds, More Especially the Falconidae', *Proceedings of the Zoological Society of London* (1912), 63–79.

Gurney, J. H. *Early Annals of Ornithology*. London: Witherby (1921).

—— 'On Flamborough Head' (1878), in *Ornithological Miscellany*, ed. G. D. Rowley, vol. 3. London: Trübner (1921), 29–38.

Hall, A. R. and M. B. Hall, eds. *The Correspondence of Henry Oldenburg*, 13 vols. Madison, WI: University of Wisconsin Press; London: Mansel; London: Taylor and Francis (1965–86).

Harvey, W. *Disputations Touching the Generation of Animals* (1653), trans. G. Whitteridge. Oxford: Blackwell (1981).

Hegenitius, G. and A. Ortelius. *Itinerarium Frisio-Hollandicum, et ... itinerarium Gallo-Brabanticum*. Leiden: Ex Officina Elzeviariana (1630).

Heneberg, P. 'On *Otis tarda* and Marquis de Sade: What Motivates Male Great Bustards to Consume Blister Beetles (Meloidae)?', *Journal of Ornithology* 157 (2016), 1,123–5.

Hoeniger, F. D. and J. F. M. Hoeniger. *The Growth of Natural History in Stuart England: From Gerard to the Royal Society*. Charlottesville, VA: University Press of Virginia (1969).

Hoffman, M. *Florae Altdorffinae deliciae hortenses sive catalogus plantarum horti Medici*. Altdorffi: Hagen (1660).

Holder, W. *A Supplement to the Philosophical Transactions of July, 1670*. London: Royal Society (1678).

Hooke, R. *Micrographia*. London: John Martyn (1665).

Howell, J. *Instructions for Forreine Travell*. London: T. B. for Humphrey Mosley (1642).

Hunter, M. *Establishing the New Science*. Woodbridge: Boydell Press (1989).

—— 'John Ray in Italy: Lost Manuscripts Rediscovered', *Notes and Records of the Royal Society of London* 68 (2014), 93–109.

—— *Science and the Shape of Orthodoxy*. Woodbridge: Boydell Press (1995).

Huynen, L., B. J. Gill, C. D. Millar and D. M. Lambert. 'Ancient DNA Reveals Extreme Egg Morphology and Nesting Behavior in New Zealand's Extinct Moa', *Proceedings of the National Academy of Sciences* 107 (2010), 16,201–6.

Iliffe, R. 'Foreign Bodies: Travel, Empire and the Early Royal Society of London. Part 1: Englishmen on Tour', *Canadian Journal of History* 33 (1998), 357–85.

Jaatinen, K., A. Lehikoinen and D. B. Lank. 'Female-biased Sex Ratios and the Proportion of Cryptic Male Morphs of Migrant Juvenile Ruffs (*Philomachus pugnax*) in Finland', *Ornis Fennica* 87 (2010), 125–34.

Jackson, C. *Great Bird Paintings of the World*, vol. 1. Woodbridge: Antique Collectors' Club (1993).

Jardine, L. *The Curious Life of Robert Hooke*. London: Harper Perennial (2003).

Jardine, W. *Nectarinidae or Sun-Birds, with a Portrait and Memoir of Francis Willughby*. London: The Naturalist's Library (1843).

Jensen, W. 'A Previously Unrecognised Portrait of Joan Baptista Van Helmont (1579–1644)', *Ambix* 51 (2004), 263–8.

Johnston, D. 'Emma Child, née Barnard, formerly Willughby (1644–1725): Records of the Life of a Gentlewoman', in *Nottinghamshire Past: Essays in Honour of Adrian Henstock*, ed. J. Beckett. Cardiff: Merton Priory Press for Nottinghamshire County Council (2003), 59–76.

—— 'The Life and Domestic Context of Francis Willughby', in *Virtuoso by Nature: The Scientific Worlds of Francis Willughby FRS (1635–1672)*, ed. T. R. Birkhead. Leiden: Brill (2016), 1–43.

Jonston, J. *Historiae naturalis de piscibus et cetis libri V.* Amsterdam: Schipper (1657).

Keynes, G. *The Works of Sir Thomas Browne*, 2nd ed. London: Faber and Faber (1964).

King, E. 'Observations on Insects Lodging Themselves in Old Willows', *Philosophical Transactions of the Royal Society* 5 (1670), 2,098–9.

Kinsky, F. C. 'The Consistent Presence of Paired Ovaries in the Kiwi (*Apteryx*) with Some Discussion of this Condition in Other Birds', *Journal für Ornithologie* 112 (1971), 334–57.

Kirkby W. *Monographia Apum Angliae*. Ipswich: Raw (1802).

Koreny, F. *Albrecht Dürer und die Tier- und Pflanzenstudien der Renaissance*. Munich: Prestel (1985).

Král, B. 'Functional Adaptations of Ciconiiformes to the Darting Stroke', *Vestnik Ceskoslovenske Spolencnosti Zoologicke, Acta. Soc. Zool. Bohemoslovenicae, Svazek* 29 (1965), 377–91.

Kusukawa, S. 'Historia Piscium (1686) and its Sources', in *Virtuoso by Nature: The Scientific Worlds of Francis Willughby FRS (1635–1672)*, ed. T. R. Birkhead. Leiden: Brill (2016), 305–34.

—— 'The Historia Piscium (1686)', *Notes and Records of the Royal Society of London* 54 (2000), 179–97.

Lack, D. *Population Studies of Birds*. Oxford: Clarendon Press (1966).

Lankester, E., ed. *Memorials of John Ray*. London: Ray Society (1846).

—— *The Correspondence of John Ray: Consisting of Selections from the Philosophical Letters Published by Dr. Derham, and Original Letters of John Ray in the Collection of the British Museum*. London: Ray Society (1848).

Latham, R. and Matthews, W. *The Diary of Samuel Pepys: A New and Complete Transcription*. Vol. 10. London: HarperCollins (1983).

Lauterborn, R., ed. *Das Vogel- Fisch- und Thierbuch des Strassburger Fischers Leonhard Baldner aus dem Jahre 1666*. Ludwigshafen am Rhein: A. Lauterborn (1903).

Leisler, B. and K. Schulze-Hagen. 'The Penetrating Gaze – New Light on Old Ideas on Mimicry and Intimidating Appearance', *Vogelwarte* 51 (2013), 55–62.

Lemery, N. *Cours de chymie*. Paris (1675).

Lewis, G. 'The Debt of John Ray and Martin Lister to Guillaume Rondelet of Montpellier', *Notes and Records of the Royal Society of London* 66 (2012), 323–39.

Lister, M. 'On a kind of viviparous fly, together with some enquiries about spiders and a table of 33 sorts to be found in England', *Philosophical Transactions of the Royal Society* 1 (1671), 600–2.

Locke, J. *An Essay Concerning Human Understanding*. London (1690).

Lokhorst, Gt.-J. 'Descartes and the Pineal Gland', *The Stanford Encyclopedia of Philosophy* (Summer 2016), ed. E. N. Zalta. http://plato. stanford.edu/archives/sum2016/entries/pineal-gland, accessed 2016.

Lownes, A. E. 'A Collection of Seventeenth-Century Drawings', *Auk* 57 (1940), 532–5.

Mabey, R. *The Cabaret of Plants*. London: Profile (2015).

MacGillivray, W. *History of British Birds, vol. 4*. London: Orr & Co. (1852).

MacPherson, H. A. *A History of Fowling*. Edinburgh: Douglas (1897).

Malina, R. 'History of Entomology and the Discovery of Insect Parasitoids', *Matthias Belvis University Proceedings* 4 (2008), 1–14.

Mandelbrote, S. 'John Ray', in *Oxford Dictionary of National Biography*, vol. 46, eds C. Matthew and B. Harrison. Oxford: Oxford University Press (2004), 178–83.

Manzini, C. *Ammaestramenti per allevare, pascere e curare gli ucceli*. Milano: Pacifico Pontio (1575).

Marcgraf, G. and W. Piso. *Historia naturalis Brasiliae*. Leiden: F. Hackium; Amsterdam: L. Elzevirium (1648).

McMahon, S. 'John Ray (1627–1705) and the Act of Uniformity 1662', *Notes and Records of the Royal Society* 54 (2000), 153–78.

Monk, J. H. and C. J. Blomfield. 'Memoir of Dr James Duport', *Museum Criticum* 2 (1826), 672–98.

Montgomerie, R. and T. R. Birkhead. 'Samuel Pepys's Hand-coloured Copy of John Ray's "The Ornithology of Francis Willughby" (1678)', *Journal of Ornithology* 150 (2009), 883–91.

Muffet, T. *Insectorum sive Minimorum Animalium Theatrum*. London (1634).

—— *Health's Improvement*. London (1655).

Mullens, W. H. 'Thomas Muffett', *Hastings and St. Leonards Natural History Society, Occasional Publication no. 5* (1911).

Mynott, J. *Winged Words: Birds in the Ancient World*. Oxford: Oxford University Press (2018).

Neri, J. *The Insect and the Image*. Minneapolis, MN: University of Minnesota Press (2011).

Newson, S. E., J. H. Marchant, G. R. Elkins and R. M. Sellers. 'The Status of Inland-breeding Great Cormorants in England', *British Birds* 100 (2007), 289–99.

Newton, A. *A Dictionary of Birds*. London: Black (1896).

Nozeman, C. *Nederlandsche vogelen volgens hunne huishouding, aert en eigenschappen beschreven*, 5 vols. Amsterdam (1770–1829).

Ogilvie, B. W. 'Stoics, Neoplatonists, Atheists, Politicians: Sources and Uses of Early Modern Jesuit Natural Theology', in *For the Sake of Learning: Essays in Honor of Anthony Grafton* 2, ed. A. Blair and A.-S. Goeing. Leiden: Brill (2016a), 761–79.

—— 'Willughby on Insects', in *Virtuoso by Nature: The Scientific Worlds of Francis Willughby FRS (1635–1672)*, ed. T. R. Birkhead. Leiden: Brill (2016), 335–59.

Olina, G. P. *Uccelliera overo discorso della natura, e proprieta di diversi uccelli*. Rome: Andrea Fei (1622).

Opie, I. and P. Opie. *The Oxford Dictionary of Nursery Rhymes*. Oxford: Oxford University Press (1951; 2nd ed. 1997).

Ostwald, W. *Grosse Männer*. Leipzig: Akademische Verlagsgesellschaft (1909).

Oswald, P. H. and C. D. Preston, eds. *John Ray's Cambridge Catalogue* (1660). London: Ray Society (2011).

Park, K. 'The Rediscovery of the Clitoris: French medicine Medicine and the Tribade 1570–1620', in *The Body in Parts: Fantasies of Corporeality in Early Modern Europe,* eds, D. Hillman and C. Mazzio. New York: Routledge (1997), 170–93.

Parker, G. A. 'Behavioral Ecology: Natural History as Science', in *Essays on Animal Behavior: Celebrating 50 Years of Animal Behavior,* eds, J. R. Lucas and L. W. Simons. Burlington, MA: Elsevier (2006), 23–56.

Phillips, J. C. 'Leonard Baldner, Seventeenth-Century Sportsman and Naturalist: An Unrecorded Copy of his Book, Containing his Portrait', *Auk* 42 (1925), 332–41.

Plot, R. *Natural History of Staffordshire*. Oxford (1686).

Poole, W. 'The Willughby Library in the Time of Francis the Naturalist', in *Virtuoso by Nature: The Scientific Worlds of Francis Willughby FRS (1635–1672)*, ed. T. R. Birkhead. Leiden: Brill (2016), 227–43.

Preston, C. D. 'Francis Willughby and Botany of the Expeditions – Evidence from Plant Specimens in the Middleton Collection', in *Virtuoso by Nature: The Scientific Worlds of Francis Willughby FRS (1635–1672)*, ed. T. R. Birkhead. Leiden: Brill (2016), 197–226.

Preston, A. and P. H. Oswald. 'James Duport's Rules for his Tutorial Pupils: A Comparison of Two Surviving Manuscripts', *Transactions of the Cambridge Bibliographical Society* 14 (2011), 317–62.

Ramsay, S. L. and D. C. Houston. 'Amino Acid Composition of Some Woodland Arthropods and its Implications for Breeding Tits and Other Passerines', *Ibis* 145 (2003), 227–32.

Ransome, A. *Great Northern?* London: Jonathan Cape (1947).

Raven, C. E. *English Naturalists from Neckham to Ray*. Cambridge: Cambridge University Press (1947).

—— *John Ray: His Life & Works*. Cambridge: Cambridge University Press (1942).

Ray, J. *Historia Piscium*. Oxford: Sheldon (1686a).

—— *Historia Plantarum*, 3 vols. London: M. Clarke or H. Faithorne (1686b, 1688, 1704).

—— *Synopsis Methodica Animalium Quadrupedum et Serpenti Generis*. London: Smith and Walford (1693).

—— *Historia insectorum: … opus posthumum Jussu Regiae Societatis Londinensis Editum. Cui subjungitur Appendix de Scarabaeis Britannicis*, Autore M. Lister. London: A. and J. Churchill (1710).

—— *Synopsis Methodica Avium & Piscium*. London: Gulielmi Innys (1713).

—— *Travels through the Low-Countries, Germany, Italy and France, with curious Observations, Natural, Topographical, Moral, Physiological, & c … To which is added, An Account of the Travels of Francis Willughby, Esq., Through great Part of Spain*, 2nd ed., 2 vols. London: Printed for J. Walthoe et al. (1738) (first edition 1673).

—— *The Wisdom of God Manifested in the Works of the Creation*. London (facsimile of 1826 edition): Ray Society Scion (2006).

—— *Further Correspondence of John Ray*, ed. R. T. Gunther. London: Ray Society (1928).

Reinertsen, R. E. 'Nocturnal Hypothermia and its Energetic Significance for Small Birds Living in the Arctic and Subarctic Regions: A Review', *Polar Research* 1 (1983), 269–84.

Reynard, M. and C. J. Savory. 'Stress-induced Oviposition Delays in Laying Hens: Duration and Consequences for Eggshell Quality', *British Poultry Science* 40 (1999), 585–91.

Romero, A. 'When Whales Became Mammals: The Scientific Journey of Cetaceans from Fish to Mammals in the History of Science', in *New Approaches to the Study of Marine Mammals*, ed. A. Romero and E. O. Keith, InTech. (2012), 3–30.

Roos, A. M. *Web of Nature: Martin Lister (1639–1712), the First Arachnologist*. Leiden: Brill (2011).

—— trans. and ed., *The Correspondence of Dr. Martin Lister (1639–1712): Vol. 1, 1662–1667*. Leiden: Brill (2015).

—— 'The chymistry of Francis Willughby (1635–72): The Trinity College, Cambridge Community', in *Virtuoso by Nature: The Scientific Worlds of Francis Willughby FRS (1635–1672)*, ed. T. R. Birkhead. Leiden: Brill (2016), 99–121.

—— '"Only meer love to learning": A Rediscovered Travel Diary of Naturalist and Collector, James Petiver (c.1665–1718)', *Journal of the History of Collections* 29 (2017), 381–94.

Sawyer, F. C. 'John Ray (1627–1705) – A Portrait in Oils', *Journal of the Society for the Bibliography of Natural History* 4 (1963), 97–9.

Schulze-Hagen, K. and T. R. Birkhead. 'The Ethology and Life History of Birds: The Forgotten Contributions of Oskar, Magdalena and Katharina Heinroth', *Journal of Ornithology* 156 (2015), 9–18.

Serjeantson, R. 'The Education of Francis Willughby', in *Virtuoso by Nature: The Scientific Worlds of Francis Willughby FRS (1635–1672)*, ed. T. R. Birkhead. Leiden: Brill (2016): 44–98.

Shrubb, M. *Feasting, Fowling and Feathers: A History of the Exploitation of Wild Birds.* Poyser; London: A&C Black (2013).

Skippon, P. 'An Account of a Journey Made thro' Part of the Low-Countries, Germany, Italy and France', in vol. 6 of *A Collection of Voyages and Travels*, ed. A. and J. Churchill. London: J. Walthoe, T. Wotton, S. Birt, D. Browne, T. Osborn, J. Shuckburgh and H. Lintot (1732), 359–736.

Smith, J. E. 'Introductory Discourse on the Rise and Progress of Natural History', *Transactions of the Linnean Society of London* 1 (1788), 1–55.

Smith, P. and P. Findlen. *Merchants and Marvels: Commerce, Science, and Art in Early Modern Europe.* London: Routledge (2001).

Storey, A. *Trinity House of Kingston upon Hull.* Hull: Trinity House (1967).

Stresemann, E. *Ornithology from Aristotle to the Present*, trans. H. J. Epstein and C. Epstein, ed. G. W. Cottrell. Cambridge, MA: Harvard University Press (1975; original edition, 1951).

—— and V. Stresemann. 'Die Mauser der Vogel', *Journal for Ornithology Sonderheft* 107 (1966).

Suh, A. 'The Phylogenetic Forest of Bird Trees Contain a Hard Polytomy at the Root of Neonaves', *Zoologica Scripta* 45 (2016), 50–62.

Swainson, W. *A Preliminary Discourse on the Study of Natural History.* The Cabinet Cyclopaedia, ed. D. Lardner, vol. 59. London: Longman, Rees, Orme, Brown, Green, and Longman (1834).

Swann, H. K. *A Dictionary of English and Folk-names of British Birds.* London: Witherby (1913).

Thomas, J. and R. Lewington. *Butterflies of Britain and Ireland*, 2nd edn. Oxford: British Wildlife (2010).

Thomas, K. 'The Life of Learning', *Proceedings of the British Academy* 117 (2002), 210–36.

Thuman, K. A., F. Widemo and S. C. Griffith. 'Condition-dependent Sex Allocation in a Lek-breeding Wader, the Ruff (*Philomachus pugnax*)', *Molecular Ecology* 12 (2003), 213–18.

Valli da Todi, A. *Il Canto de gl'Augelli*. Rome: N. Mutij (1601).

Vestjens, W. J. M. 'Notes on the Trachea and Bronchi of Australian Spoonbills', *Emu* 75 (1975), 87–8.

Vogelsang, H. *Der Behemer (bergfink: Fringilla montifringilla)*. Bergzabern: Pfeifer u Messbecher (1949).

Wardhaugh, B. 'Willughby's Mathematics', in *Virtuoso by Nature: The Scientific Worlds of Francis Willughby FRS (1635–1672)*, ed. T. R. Birkhead. Leiden: Brill (2016), 122–41.

Welch, M. A. 'Francis Willoughby, F.R.S. (1635–1672)', *Journal of the Society for the Bibliography of Natural History* 6 (1972), 71–85.

Whitten, A. J. 'A New Behavioural Method for Further Determination of Olfaction in Mallard (*Anas platyrhynchos*)', *Journal of Biological Education* 5 (1971), 291–4.

Wilkins, J. *An Essay towards a Real Character, and a Philosophical Language*. London: Sa. Gellibrand and John Martyn (1668).

Williams, G. C. *Adaptation and Natural Selection*. Princeton, NJ: Princeton University Press (1966).

Willughby, F. 'Extracts of two letters, written by Francis Willoughby Esquire, to the publisher, from Astrop, August 19th, and from Midleton, Sept. 2d. 1670 containing his observations on the insects and cartrages, described in the precedent accompt', *Philosophical Transactions of the Royal Society* 5 (1670), 2100–02.

—— 'Observations about that kind of wasps called Vespae ichneumons', *Philosophical Transactions of the Royal Society* 6 (1671), 2279–81.

Wood, A. C., ed., *The Continuation of the History of the Willughby Family by Cassandra Duchess of Chandos*. Windsor: Shakespeare Head Press for the University of Nottingham (1958).

Wood, C. T. *The Ornithological Guide*. London: Whittaker (1835).

Wood, N. *The Ornithologist's Text-Book*. London: John W. Parker (1836).

Woodward, J. and D. Goodstein. 'Conduct, Misconduct and the Structure of Science', *American Scientist* 84 (1996), 479–90.

Wootton, D. *The Invention of Science: A New History of the Scientific Revolution*. New York: Harper Collins (2015).

Wragge-Morley, A. 'The Work of Verbal Picturing for John Ray and Some of his Contemporaries', *Intellectual History Review* 20 (2010), 165–79.

Yeo, R. *Notebooks, English Virtuosi, and Early Modern Science*. Chicago, IL: University of Chicago Press (2014).

Zeiler, M. *Hispaniae et Lusitaniae Itinerarium*. Amsterdam: Janonsius Valckenier (1656).

Zimmerman, M. B. 'Research on Iodine Deficiency and Goiter in the 19th and early 20th Centuries', *Journal of Nutrition* 138 (2008), 2,060–3.

Zwinger, T. *Methodus apodemica in eorum gratiam qui cum fructu in quocunque tandem vitae genere peregrinari cupiunt*. Basel: Eusebii Episcopii (1577).

Picture Credits

Note: M&SCUN refers to Manuscripts and Special Collections, University of Nottingham

pp. 9, 133: Wikimedia Commons

pp. 17, 53, 120, 245, 256 (left): M&SCUN, courtesy Lord Middleton

p. 24: From the author's personal copy of Ray (1673); this image (made later) has been tipped in.

pp. 48, 80: Illustration by David Quinn

p. 51: Gerarde's *Herball* (1597)

p. 52: Plot (1686)

pp. 59, 84, 104, 123, 145: Redrawn from Greengrass et al. (2016)

pp. 65, 69: By kind permission of Chapter of Gloucester Cathedral

p. 67: Redrawn from Carter & Basire (1807)

p. 86: From Cornelius Nozeman's five-volume *Nederlandsche Vogelen* (1770–1829), courtesy Artis Library, University of Amsterdam

p. 88: Aldrovandi (1603), vol III, p. 387

pp. 95, 233, 240: Ray (1678)

p. 112: Aldrovandi (1648)

p. 129: Jonston (1657)

p. 141: Olina (1622)

pp. 151, 162: Wilkins (1668)

p. 157: Willughby (1670), courtesy the Royal Society of London

p. 164: Courtesy Lord Middleton

p. 177: Courtesy the Royal Society of London

p. 182: Courtesy Haddon Hall Estates

p. 184: Cram et al. (2003)
p. 207: van Grouw (2103)
p. 218: Charleton (1668)
p. 231: Belon (1555)
p. 245 (right): Ray (1686)

Colour plates

Section 1
Portraits of Francis Willughby and his parents: courtesy © Birdsall
    Estates
Middleton Hall: photo by D. Johnston
Green and great spotted woodpeckers: © Middleton Collection
    MiLM24/26, 105
Black-eared wheatear, lesser grey shrike and red kite: © Middleton
    Collection MiLM24/39, 66, 31
Night heron, perch and insects: © Brown University
Three fish: © Middleton Collection MiLM 25/77, 78, 55

Section 2
Spiky beetle, eggs, curiosities and seed trays: photos by T. R.
    Birkhead, courtesy © Birdsall Estates
Gecko and dormouse: © Middleton Collection MiLM24/167, 163
Leaf-cutter bee: photo © P. Donald
Bee cartrages: photo © J. Early
Charr: courtesy © Zoological Society of London
Dried plant specimens: © Middleton Collection MiML20
Honey-buzzard: photo by K. Nigge

Section 3
Red-backed shrike: photo by T. R. Birkhead
Pigeons: https://www.flickr.com/photos/nottsexminer/8012803261/
in/photostream/
European roller: photos by S. Butler
Eurasian bittern: photos by T. R. Birkhead
Pigo: photo by A. Hartl

Index

Page numbers in **bold** refer to illustrations

A Note on the Author

Tim Birkhead FRS is Professor of Zoology at the University of Sheffield, where he teaches animal behaviour and history of science. Among his other books are *Promiscuity*; *Great Auk Islands*; *The Cambridge Encyclopedia of Birds*, which won the McColvin medal; *The Red Canary*, which won the Consul Cremer Prize; The Wisdom of Birds, which was the *British Birds*/British Trust for Ornithology Best Bird Book of the Year 2009; *Bird Sense*, which was the *Guardian* and the *Independent*'s Natural History Book of the Year; and *The Most Perfect Thing*, winner of the Zoological Society of London's Communicating Zoology Award for 2016. He lives in Sheffield.

A Note on the Type

The text of this book is set Adobe Garamond. It is one of several versions of Garamond based on the designs of Claude Garamond. It is thought that Garamond based his font on Bembo, cut in 1495 by Francesco Griffo in collaboration with the Italian printer Aldus Manutius. Garamond types were first used in books printed in Paris around 1532. Many of the present-day versions of this type are based on the *Typi Academiae* of Jean Jannon cut in Sedan in 1615.

Claude Garamond was born in Paris in 1480. He learned how to cut type from his father and by the age of fifteen he was able to fashion steel punches the size of a pica with great precision. At the age of sixty he was commissioned by King Francis I to design a Greek alphabet, and for this he was given the honourable title of royal type founder. He died in 1561.